THE GEEK MANIFESTO

Why Science Matters

MARK HENDERSON

BANTAM PRESS

LONDON · TORONTO · SYDNEY · AUCKLAND · JOHANNESBURG

TRANSWORLD PUBLISHERS
61–63 Uxbridge Road, London W5 5SA
A Random House Group Company
www.transworldbooks.co.uk

First published in Great Britain
in 2012 by Bantam Press
an imprint of Transworld Publishers

A CIP catalogue record for this book
is available from the British Library.

ISBNs 9780593068236 (hb)
9780593068243 (tpb)

Addresses for Random House Group Ltd companies outside the UK
can be found at: www.randomhouse.co.uk
The Random House Group Ltd Reg. No. 954009

The Random House Group Limited supports the Forest Stewardship
Council (FSC®), the leading international forest-certification organization.
Our books carrying the FSC label are printed on FSC®-certified paper. FSC
is the only forest-certification scheme endorsed by the leading environmental
organizations, including Greenpeace. Our paper procurement policy can be
found at www.randomhouse.co.uk/environment.

Typeset in 11/15pt Times by Falcon Oast Graphic Art Ltd.
Printed and bound in Great Britain by
Clays Ltd, Bungay, Suffolk

2 4 6 8 10 9 7 5 3 1

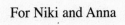

For Niki and Anna

Contents

'We've arranged a global civilization in which most crucial elements – transportation, communications, and all other industries; agriculture, medicine, education, entertainment, protecting the environment; even the key democratic institution of voting – profoundly depend on science and technology. We have also arranged things so almost nobody understands science and technology. This is a prescription for disaster. We might get away with it for a while, but sooner or later this combustible mixture of ignorance and power is going to blow up in our faces.'

Carl Sagan

'A habit of basing convictions upon evidence, and of giving to them only that degree of certainty which the evidence warrants, would, if it became general, cure most of the ills from which this world is suffering.'

Bertrand Russell

THE GEEKS
ARE COMING

IT WAS CHIROPRACTIC AWARENESS WEEK. BUT THE AWARENESS SIMON Singh was raising wasn't quite what the British Chiropractic Association had in mind.

'You might think that modern chiropractors restrict themselves to treating back problems,' the science writer declared in the *Guardian* on 19 April 2008, 'but in fact they still possess some quite wacky ideas. The British Chiropractic Association [BCA] claims that their members can help treat children with colic, sleeping and feeding problems, frequent ear infections, asthma and prolonged crying, even though there is not a jot of evidence. This organization is the respectable face of the chiropractic profession and yet it happily promotes bogus treatments.'

The notion that cracking a little girl's spine might cure her earache indeed sounds curious, but plenty of hypotheses that were once dismissed as wacky have been accepted as science on the back of sound data. The respectable face of chiropractic could thus have defended itself with respectable evidence. The BCA, which had none, chose another way to fight back. It sued Singh for libel.

That didn't look a bad idea at the time. England's defamation laws are notoriously friendly to plaintiffs, who need not show that an alleged libel is false. The burden of proof lies on the defendant, who must demonstrate his assertion to be true, or a matter of fair comment. The legal costs involved are so steep, often running to hundreds of thousands of pounds, that even defendants who are sure of their facts can be cowed into submission for fear of bankruptcy. With the financial resources to sue, the BCA saw an opportunity to force one of Britain's most vocal and effective critics of alternative medicine to eat his words.

The chiropractors, however, hadn't reckoned on the geeks.

It was an easy mistake to make. People with a passion for science and the critical thinking on which it is founded have never been particularly conspicuous in public life, let alone formed a constituency to be crossed at your peril. Yet something is stirring among those curious kids who always preferred sci-fi to celebrity magazines and chemistry sets to trendy trainers. We've stopped apologizing for our obsession with asking how and why, and we're starting to stand up for ourselves instead.

Drawn together by the social networking power of the internet, geeks have begun to realize that we aren't alone in our world view, but that it's shared by millions. Through vibrant blogs and online forums such as Twitter and Facebook, and through the success of increasingly high-profile figures such as Singh, Brian Cox and Ben Goldacre, we've begun to fight for the value we place on science and evidence-based thinking.

What was once a term of derision has been embraced as a badge of honour, in a surge of geek pride. Armed with new confidence, and the online means to discover one another and spread our message, we are finding a public voice that is proving more powerful than we could possibly have imagined.

The geeks are on the march. The BCA foolishly threw itself in the way.

Even before the chiropractors decided to mess with one of our number, many geeks had become dogged pursuers of homeopaths, anti-vaccine activists, dodgy nutritionists and other purveyors of quackery and pseudoscience. Some geeks are scientists. Some are doctors. Many are neither. All, though, care deeply about the scientific method: the most reliable tool humanity has yet developed for distinguishing truth from falsehood. 'We're rationalists,' as Singh puts it. 'We aren't necessarily scientists, but we have an affinity for science.'

Geeks take a forensic approach to the evidence behind medical claims, and are strongly committed to unfettered debate. The chiropractors' writ could scarcely have been better calculated to rile us. Legal bullying was shutting down rational argument. No self-respecting geek was going to stand for it.

As news of the lawsuit reached the blogosphere, geeks bearing *noms de plume* such as Gimpy and Zeno, the Quackometer and Adventures in Nonsense rallied to Singh's support. Almost 10,000 people joined a Facebook group started by David Allen Green, a lawyer who blogs as Jack of Kent. Others weighed in on Twitter. Many offered money to finance the defence, which Singh declined. As the bestselling author of *Fermat's Last Theorem*, *Big Bang* and *The Code Book*, he had the means to fight. The question was whether he had the stomach for a legal battle that might effectively become a full-time job for years on end.

It was a battle, too, which evidence alone might be insufficient to win. In May 2009, Mr Justice Eady ruled that, by calling their claims 'bogus', Singh had accused the chiropractors of deliberate dishonesty, an implication he never intended and could not really defend. Yet as he weighed up whether or not to settle and apologize, the groundswell of geek support steeled his nerve.

Quacklash

A few days after Mr Justice Eady's 'bogus' ruling, the Penderel's Oak pub in Holborn, central London, thronged with geeks. Summoned by blogs, Twitter and Facebook, they had come to discuss Singh's options. As supporters who had never met the science writer – let alone one another – shared their indignation, the mood of the meeting grew defiant.

'The reaction was extraordinary,' Singh says. 'There was a point in May when I was close to caving in. That support was really important. It made me think: "Simon, you're not crazy. You're not the only one who thinks this matters."'

Green agrees that the geeks were crucial. 'No-one would have thought badly of Simon if he had just brought the case to a halt,' he wrote later on his blog. 'The ever-growing online support helped keep him soldiering on.'

It wasn't just moral support that the geeks had to offer. A devastating counter-attack was soon under way. When the BCA released what it called a 'plethora of evidence' supporting chiropractic as an effective treatment for childhood ailments such as colic and asthma, a battalion of bloggers demolished every claim within twenty-four hours. If the plaintiffs were to rely on this in court, the defence would have refutations to hand.

Then there was what Green dubbed the 'quacklash'.

Unlike most other alternative therapists, such as homeopaths or reflexologists, chiropractors are regulated in the UK. They must adhere to a set of professional guidelines that include obtaining informed consent from their patients, and they are subject to trading and advertising standards, which do not allow claims that are not supported by evidence.

Bloggers such as Andy Lewis (*the Quackometer*), Simon Perry (*Adventures in Nonsense*) and Alan Henness (*Zeno's Blog*) began to trawl chiropractors' websites for misleading and unsupported medical assertions. They then reported those who appeared to be in breach of regulatory standards. 'I don't think there could be a better use of £75 worth of stamps,' wrote Perry.

There was no shortage of suitable targets. In June 2009, shortly after Mr Justice Eady's preliminary ruling, the General Chiropractic Council received complaints about more than 500 individual practitioners in just twenty-four hours. Chiropractors went into full-blown damage-limitation mode. Lewis got hold of an email from the McTimoney Chiropractic Association urging its members to take down their websites and 'to remove any patient information leaflets of your own that state you treat whiplash, colic or other childhood problems in your clinic'. These, of course, were the very claims Singh had questioned to prompt the BCA's writ.

By resorting to law, the back-crackers inflicted terrible self-harm. In trying to silence a critic, the BCA invited unprecedented scrutiny of the evidence base for its techniques. Newspapers that hadn't covered the original lawsuit gleefully reported the quacklash and the threat to free speech. There was suddenly a news hook for articles examining the questionable claims made for chiropractic, and the Kafkaesque anachronisms of English libel law.

Even were the BCA to win at trial, damage to the reputation it sued to protect would reach a different scale to anything inflicted by Singh's original column. But the victory was to be Singh's. In April 2010, the Court of Appeal overturned Mr Justice Eady's ruling in a withering judgment, and the chiropractors dropped their case. Singh's supporters celebrated on their blogs and on Twitter by posting: 'The BCA happily promotes bogus treatments.' One in four British chiropractors was under investigation by regulators at the time.

Geek activism had helped Singh to win a seminal case, which established an important legal precedent that should protect other scientists and writers. 'Scientific controversies must be settled by the methods of science rather than by the methods of litigation,' the judgment noted. But the campaign achieved something else besides, focusing public attention on the chilling effect of English libel law on public discourse.

The libel action turned Singh – well known for a science writer but hardly a household name – into a cause *célèbre*. He became a symbol of free speech and principled skepticism,[1] championed by celebrities such as Ricky Gervais and Stephen Fry. Galvanized by the packed public meeting at the Penderel's Oak, Tracey Brown and Sile Lane, of the charity Sense About Science, began a 'Keep Libel Laws Out of Science' petition, soon to carry 20,000 names. Out of that grew a wider campaign for libel reform which is now starting to bear fruit.

The narrow aversion of a serious injustice, which still left a vindicated Singh £60,000 out of pocket because of legal costs he could not recover, offered convincing evidence that libel law had become a serious threat to free expression. Labour, the Liberal

[1] British geeks generally prefer the American spelling 'skeptic' and 'skepticism', with a k instead of a c, when it's used to describe an allegiance to rationalism and science. This distinguishes such skeptics from sceptics who don't really found their doubts on science, such as climate sceptics.

the Conservatives all pledged to support reform

0 election manifestos. In March 2011, the

–Liberal Democrat coalition introduced a draft

ill, which will create important new defences of public

honest opinion. The final version will be published at

about̸ ame time as this book.

Journalists, human rights activists and some lawyers had been complaining for years about the iniquities of English libel law, to little effect. A bunch of geeks, led by a brave and resourceful figure-head in Singh, provided the catalyst for change.

The value of science

The Singh case was a victory for an emerging force with the potential to change politics and society for the better, which has slowly begun to gather strength over the past half decade. Fed up with being marginalized, and mustered by new media, the geeks are coming. Our countries need us.

Libel law is far from the only arena of public life that could use our influence. The majority of the most pressing contemporary problems would be a little more tractable if our leaders were to listen more care-fully to what geeks have to say. From drugs to climate change, from education to the economy, the scientific approach that we champion has more to contribute to effective public policy than most politicians and civil servants have hitherto acknowledged.

As those of us who care deeply about science and its experimental method start to fight for our beliefs, geeks have a historic opportunity to embed critical thinking more deeply in the political process. But if we are to achieve anything, we need to turn our numbers and confidence into political muscle.

The Geek Manifesto seeks to define this challenge and to suggest how we can rise to it. It will explain how and why politics lets science down and fails to exploit its powerful approach to evidence in pursuit of effective policy. Above all, it will explore how geeks can turn our irrepressible energy and analytical rigour into a movement with real clout – how we can move on from railing against science abuse and begin to prevent it.

This isn't a traditional manifesto in the sense of a laundry list of policy prescriptions. When it does make detailed proposals, it is on the understanding that not all geeks will agree. Science thrives because it is always open to new ideas, so long as they can survive skeptical scrutiny, and this burgeoning movement should likewise draw strength from diversity and criticism. The manifesto's aim is to win your broad support for its central proposition: that a more scientific approach to problem-solving is applicable to a surprisingly wide range of political issues, and that ignoring it disadvantages us all. Precisely what politicians think is less important than how they think.

Precisely what politicians think is less important than how they think

Most of our leaders believe science is one of two things. It is a collection of facts, a body of knowledge about the world that can be taught and learned, such as gravity, evolution or photosynthesis. Or it is the technology that is the fruit of that knowledge: the computers, vaccines and aeroplanes that change the way we live.

Neither is wrong, but as geeks appreciate, science is something else besides. 'Science is more than a body of knowledge,' said Carl Sagan, the great astronomer and popularizer of science. 'It is a way of thinking.' In the metaphor of the skeptical writer Michael Shermer, it is not a noun but a verb. It is something people do, a method like no other for establishing how the world works. You have an idea, then you gather the soundest possible evidence against which to evaluate it. If the data allow, your idea can be tentatively accepted. If not, it must be set aside.

Science is provisional, always open to revision in the light of new evidence: it is comfortable with changing your mind, indeed it often insists on it. It is anti-authoritarian: anybody can contribute, and anybody can be wrong. It makes testable predictions and then seeks actively to test them. Over time, it is self-correcting, because of the importance it places on trying to prove the most elegant ideas wrong. It is comfortable with uncertainty, knowing that even its best answers will simply be better approximations of the truth.

This scientific method is not perfect by any stretch: scholars of 'science studies' have shown effectively that the idealized picture presented in the last paragraph is rarely quite fulfilled in practice. Individual scientists are people, with prejudices, motives and values, and they make mistakes. What sets their approach to thinking apart is that they are aware of these limitations, and try to compensate for them by taking a skeptical attitude even to cherished and beautiful ideas – not least because they appreciate that if they fail to test them properly, somebody else certainly will. 'Science is a way of trying not to fool yourself,' explained Richard Feynman, the Nobel prize-winning physicist. 'The first principle is that you must not fool yourself, and you are the easiest person to fool.'

This embedded self-criticism makes science, to paraphrase Sir Winston Churchill, the worst means of discovery, excepting all the others that have been tried from time to time. Its power, however, is too often confined to an intellectual ghetto: those disciplines that have historically been considered to be 'scientific'. Though this method of inquiry has great things to contribute to all sorts of pursuits beyond the laboratory, it remains missing in action from far too much of public life.

Its absence matters more than ever because, as John Holdren, President Obama's chief science adviser, explains: 'more and more of the public policy issues that are before [us] . . . have science and technology content'. If we are to stand a chance of containing global warming, the solutions will rely on understanding the atmospheric effects of greenhouse gases, the technologies that can generate clean energy and the psychology of behaviour change. If we are properly to exploit advances in genetics and neuroscience to deliver better healthcare, and to spend national resources on the drugs and therapies that are most beneficial to patients, science will be central.

As a source of innovation and business ideas, science is a foundation stone of growth. It is vital to economies like Britain's and America's, which cannot compete on labour costs with China, India and Brazil. Food security, drugs control, forensic investigations, pandemics: the list of policy questions that require politicians to be intelligent consumers of science is as long as your arm.

It's actually hard to think of an issue to which science is irrelevant. When Sir David King was the chief scientific adviser to Tony Blair's government, he challenged senior civil servants to suggest an area of public policy in which science didn't matter. One mandarin from the Department for Work and Pensions piped up to claim that his whole portfolio could do without it. The challenges this department faces over the next half century will be defined by Britain's ageing population. Yet a senior official within it felt that science had nothing worthwhile to contribute.

Science matters when it has amassed good evidence that defines a policy challenge, such as climate change, and the effects that different solutions might have. In many ways, it matters still more when the evidence base is weak and questions are more open. If ministers want to know how best to teach children to read, or how best to rehabilitate drugs offenders, they can use the methods of science to find out.

Even when a challenge is so urgent that there is no time to wait for research, ministers can at least ensure that the policy solution is properly evaluated so that lessons can be learned from its successes and shortcomings. Policy decisions aren't the last word, but they are the start of experiments that could and should be mined for evidence that can be used to make better choices in future.

It's true that you can't study schools or prisons by setting up a controlled experiment in a laboratory, but neither can you recreate the Big Bang or re-run human evolution in a test tube to watch how they happened. Politics asks tough questions of our leaders, but science provides a great tool for answering them. If it can help us to understand the first microseconds of creation and the descent of man, the scientific method can surely improve understanding of how best to tackle the pressing social questions of our time. Teaching techniques, sentencing policies, policing strategies: all could be investigated with this tool to establish whether they work. All too often they are not, leaving everybody uncertain about their worth.

This value of science, however, is seldom grasped either by the ministers, advisers and officials who take the decisions that shape everybody's lives or by opinion-formers in the media and think-tanks to whom they typically listen. The privileged place of the humanities and their graduates in politics, the media and public intellectual life, famously identified by C. P. Snow in his 'Two Cultures' lecture of 1959, is still a serious issue today. There are fifty-five US senators with law degrees – more than half the Senate's membership. None has a PhD in the natural sciences, and only one, Chris Coons of Delaware, has an undergraduate science degree. The UK's 650 MPs include just three with science PhDs, only one of whom, Julian Huppert of Cambridge, has worked in research.

You don't have to be a trained scientist to understand what science has to offer. Several of the most effective recent British science ministers, such as Lord Sainsbury of Turville and the present in-cumbent, David Willetts, are humanities graduates, as are backbench champions such as Phil Willis, who adeptly chaired the House of Commons Science and Technology Select Committee for many years. Their presence, though, is insufficient to compensate for heavy under-representation of one of the most important professions in our society.

Politics suffers accordingly. As we will see, too few of our leaders understand either the conditions science needs to thrive or the powerful contribution it can make to policy-making. They see science as an optional extra, to be used when it suits an agenda and to be ignored when it does not. And they do this in large part because we let them. Science isn't a voting issue. Abusing and undervaluing it carries no political cost.

It doesn't have to be that way. Fortunately, there's an increasingly vibrant community out there with the potential to form that constituency and to create that political cost. I'm talking about the geeks.

The age of the geek

There has never been a better time to be a geek. Poorly served for years by the mainstream media and entertainment, we are demanding, and getting, more and more attention – in popular culture if not yet in politics. Proper geeks are becoming proper celebrities, with an impact and reach that stretches well beyond their core audiences. Whisper it, but being a geek is becoming cool.

Professor Brian Cox has brought rock-star cool to physics, turning entropy and relativity into mainstream Sunday-evening television entertainment. His *Wonders of the Solar System* and *Wonders of the Universe* programmes have topped BBC2's ratings with well over 3 million weekly viewers. His sex appeal, irreverence and boyish enthusiasm have made him a coveted guest on popular talk shows that allow him to reach a completely new audience. Jonathan Ross even presented him with a model of the Large Hadron Collider made of sex toys, bringing new meaning to his catchphrase 'And that's why I love physics.'

Cox is a geek who wears the label with pride. Many people know that he used to play keyboards in D:Ream, the band whose anthem 'Things Can Only Get Better' was memorably used by New Labour in the 1997 general election. Fewer know that he was a teenage bus-spotter. 'I liked the Class 51 that Greater Manchester used to have,' he told Hugo Rifkind of *The Times*. 'I used to write down the numbers. But I went from that to being basically, well, going to clubs and listening to music, and being in a band within a year. But I was always into the science. It was always a part of my psyche.'

Ben Goldacre is another self-proclaimed geek who has become a champion of his tribe. The targets of his 'Bad Science' column in the *Guardian* – the mail-order PhD nutritionist Gillian McKeith, manipulative drug companies and scaremongering journalists – have enraged geeks for years. Goldacre gave this outrage a media voice. His *Bad Science* book was a bestseller, shifting more than 300,000 copies in the UK alone. Goldacre's take-no-prisoners approach has inspired hundreds of other geeks to start blogging in similar fashion, exposing quackery and pseudoscience wherever they find it. Anybody who plays fast and loose with the evidence must now live in fear of being 'Goldacred' – whether by Goldacre himself or by one of his many confederates.

On stage and television, many of Britain's most successful comedians confess to geeky stirrings. Dara O'Briain is a physics graduate who peppers his routine with jokes about homeopathy. Ben Miller, of the Armstrong and Miller double act, started a PhD in quantum physics before his comedy career took off. Tim Minchin, the comic pianist, performs a nine-minute beat poem called 'Storm': a diatribe aimed at a new-ager spouting pseudoscientific nonsense at a chattering-class dinner party.

Then there's Robin Ince, who as well as reading Feynman and Sagan in his gigs – and describing Simon Singh's antagonists as 'spine wizards' – has become a geek impresario. He co-hosts with Cox *The Infinite Monkey Cage*, the award-winning Radio 4 programme that blends science with comedy, and has brought a similar fusion to the stage. Ince's *The School for Gifted Children*, *Nine Lessons and Carols for Godless People* and *Uncaged Monkeys* mix comedy from the likes of O'Briain and Minchin with short, sharp, thought-provoking presentations by Cox, Singh and Richard Dawkins. In December 2009, a capacity crowd of 3,600 filled the Hammersmith Apollo for the *Nine Lessons*. O'Briain opened his headline slot with a cry of: 'Welcome to Nerdstock!'

The success of Nerdstock caught Ince completely by surprise. 'People were waiting for something like this,' he says. 'It's all about what I'd call the rise of the new geeks. The shame is disappearing. It's been building for ten years. People like Brian and Ben are proper geeks, but they're cool enough to have that wider appeal. Brian might be a polished presenter, but start talking to him about physics or science funding, and he's a proper geek all right.'

It was the enthusiasm of confirmed geeks that made these shows work in the first place, but they're now breaking out beyond this core market. 'It's about reigniting a passion for science that might have got lost somewhere in people's secondary education,' says Ince. 'I get people coming up to me after gigs to ask what they should read by Sagan and Feynman. I get mums saying they listen to the *Monkey Cage* with their children on the way back from school. This stuff reaches a huge audience, which might not get it any other way.'

A growing public appetite for live science doesn't end with the funny stuff. Science festivals are going from strength to strength. The Cheltenham Science Festival, first staged in 2002, sold more than 30,000 tickets in 2011, with another 12,000 people attending free events. Cambridge, Nottingham, Edinburgh and Brighton are just a few of the cities that stage similar events. New York's World Science Festival, founded in 2008, attracts hundreds of thousands of people. The Amazing Meeting, an annual celebration of science and skepticism founded by the magician and quackbuster James Randi, sells out in hours in both Britain and the United States.

The mainstream media has started to respond to this resurgent public demand for science. In 2009, *The Times* launched *Eureka*, a successful monthly science magazine, reasoning that people who want intelligent, in-depth coverage of science have been poorly served by newspapers over the years. The BBC made 2010 its 'Year of Science'. The *Guardian* has established a science blogging network. Everybody wants a piece of the geeks.

Online social networking has allowed geeks to find more people of like mind, and to meet each other in the flesh and form real communities. Dozens of British towns and cities now have branches of Skeptics in the Pub (SiTP), which hold debates and discussions about science and critical thinking. 'There is an ever expanding army of geeks and the wonderful thing about it is its somewhat anarchic nature,' says David Colquhoun, Professor of Pharmacology at University College London and an SiTP regular. 'No organization, no hierarchy, just a collection of people with similar ideas, trying to improve things in their spare time. It doesn't matter how old you are (15 to, ahem, 75), or how many letters you have after your name, you can take part on equal footing.'

There is a huge community out there with a strong affinity for things scientific and an interest in contributing to the world. 'We are entering the age of the geek,' Cox told the *Sunday Times*. 'The science scene – or maybe we should call it the rational thought scene – is definitely developing. It's cool these days to actually think. My optimistic hope is that it will become very cool to really think about things . . . rather than do reactive bullshit based on no knowledge.'

A political challenge

If we want to make sure things can only get better, this growing cultural confidence isn't enough. The geek awakening needs to spawn a political movement – a popular force that nobody running for office feels they can safely ignore, and that the shrewder ones will want to appeal to and exploit.

There is an analogy here, if an imperfect one, with what groups like Stonewall have done for gay rights. A respect for science is obviously very different from sexual orientation, and we shouldn't imagine that geeks are the victims of discrimination and hate in the same way that homosexuals have been. But gay politics has none the less achieved something important that we can seek to emulate. It has changed perceptions, making casual homophobia or even indifference to gay rights unacceptable among people who seek office in Britain. Politicians know that if they fail to engage with the gay community, and fail to develop coherent positions on the issues that concern them, they risk punishment at the ballot box.

If we can do that for science, we will have made an outstanding start. Too few people who care about science make it a deciding factor at the ballot box. Geeks, like gays, will always weigh all sorts of issues when we vote. But science must be one of them.

The numbers are with us. The Campaign for Science and Engineering estimates that more than 3 million people in Britain have some sort of science background: a relevant degree, or a job in a research-intensive industry such as pharmaceuticals or IT. That amounts to about 7 per cent of the electorate, or almost as many voters as all the ethnic minorities put together. In the US, the National Science Foundation counts at least 5.5 million working scientists and engineers, to say nothing of others with a broader scientific education or career. Add in their families, and the many lay people who value science, and there is a vast constituency waiting to be tapped.

We're moving in the right direction. Geek activism helped to swing the Singh case and to force libel reform on to the political agenda. When UK science funding was threatened in the 2010 spending review, clever lobbying and a grassroots petition turned mooted cuts of 30 per cent into a spending freeze. When Professor David Nutt was sacked in 2009 as the government's chief drugs adviser for questioning evidence for classification decisions on cannabis and ecstasy, ministers were taken aback by the scale of the geek backlash and changed the rules by which they consider scientific advice. An appetite for political science is there.

These have been real achievements, which demonstrate how geeky voices, raised in the right way, can make a difference. But there is so much more to be done. If geeks are finally starting to show up on politicians' radar, we remain little more than a blip. We must learn from these campaigns, and take them much further.

There's a real opportunity here. If we get things right, we have the chance to create a constituency that politicians not only have to notice, but one they actually want to attract. Before the 2010 general election, David Cameron, Gordon Brown and Nick Clegg did everything they could to engage with Mumsnet, the spectacularly popular and increasingly influential website for mothers. Its success is a good model for geeks: mums don't agree on everything in politics any more than geeks do, but they have a set of broadly common values. It isn't too much to hope that future candidates for prime minister or president might be equally keen to court the geek vote.

Let's make that happen. Let's create a political cost for failing science. Politics has had it too easy for too long. It's time for a geek revolution.

GEEKING THE VOTE

Why science matters to politics

ON THE EVENING OF 14 OCTOBER 2009, DAVID TREDINNICK GOT TO HIS feet in the House of Commons to open a debate. The Conservative MP for Bosworth, in Leicestershire, was desperately worried that the Health Minister, Gillian Merron, had overlooked a grave threat to the wellbeing of the public. The object of his concern wasn't pandemic flu, or air pollution, or childhood obesity. It was the moon.

'At certain phases of the moon there are more accidents,' he gravely informed the House. 'Surgeons will not operate because blood clotting is not effective, and the police have to put more people on the streets.' It wasn't the first time he had raised the subject in Parliament. Back in 2001, Tredinnick told the Commons that 'science has worked out that pregnancy, hangovers and visits to one's GP may be affected by the awesome power of the moon', and quoted a newspaper report suggesting that arson attacks double when the moon is full.

He stopped short of mentioning werewolves, but you probably don't need to be told that their existence is about as well supported by science as his other claims.

So convinced is Tredinnick of the political significance of the movements of the heavens that he charged the taxpayer £755.33 for astrology software and consultancy services (which he later repaid when his expense claim became public). His commitment to the lunatic fringe of science does not end there: he is an assiduous promoter of just about every alternative medicine on the market, and recently asked the Health Secretary to congratulate homeopathic chemists on their contribution to containing swine flu.

It's tempting to think of Tredinnick as little more than a harmless eccentric, with opinions so far outside the mainstream that they carry very little influence. Would that this were so. In the summer of 2010, his fellow Conservative MPs elected him to a seat on the House of Commons Health Select Committee. Yes, a man who genuinely appears to believe that surgeons prefer not to operate when the moon is full, and who has called on the Department of Health to be 'very open to the idea of energy transfers and the people who work in that sphere', is now among the eleven politicians tasked with holding that department to account.

He isn't alone. Serving alongside Tredinnick on the health committee we find Nadine Dorries, a Tory MP who likes to promote an urban myth about a twenty-one-week foetus grasping a surgeon's finger – repeatedly denied by the surgeon – to support her demand for restricting abortion.

Neither is a fondness for pseudoscience confined to the backbenches. Peter Hain, a long-serving minister in the governments of Tony Blair and Gordon Brown, convinced himself that homeopathy cured his son's eczema, and promoted alternative medicine from a position of power. Anne Milton, a current Conservative Health Minister, cites her grandmother's experience as a homeopathic nurse in support of NHS funding of alternative medicine.

In the United States, weird views about the findings and importance of science straddle party boundaries in similar fashion. At a House of Representatives hearing in 2007 on the Intergovernmental Panel on Climate Change's most recent report, Dana Rohrabacher, a Republican congressman from California, took issue with conventional explanations for sharp global warming in prehistoric times. 'We don't know what these other cycles were caused by in the past,' he said. 'Could be dinosaur flatulence, you know, or who knows?' When Tom Coburn, an Oklahoma senator and medical doctor, asserted that 'condoms do not prevent most STDs', his reward was to be appointed by President Bush to the chairmanship of an HIV advisory group.

Tom Harkin, the influential Democratic senator for Iowa, convinced that his allergies were cured by a supplement known as bee pollen, secured the creation of the US National Center for Complementary and Alternative Medicine, which wastes about $130 million a year on studies of what you might call bogus therapies. Gary Goodyear, Canada's Science Minister, is a chiropractor who in 2009 refused to say whether he believed in evolution, telling a journalist from Toronto's *Globe and Mail*: 'I'm not going to answer that question. I am a Christian, and I don't think anybody asking a question about my religion is appropriate.'

Houses of indifference

We'll meet many of these characters again later, in Chapters 8 and 9. But, thankfully, they're not all that typical of politicians. The level of scientific misunderstanding they show, which sometimes borders on outright hostility to science and its methods, lies at the extreme end of the spectrum. Yet the very fact that they have been able to succeed in politics despite such opinions, and to rise to positions of considerable power and influence, is indicative of the value that politics places on science. Too few politicians even recognize the absurdity of their views. Tredinnick and Dorries aren't figures of fun who lack the respect of their colleagues: they were *elected* to the select committee where their unscientific positions have the potential to do most damage.

If mercifully few politicians are actively anti-science, many are indifferent to it. They often lack an understanding and appreciation both of basic scientific concepts and language and, more importantly, of its robust approach to developing reliable knowledge. Many are simply uninterested. In the last House of Commons the Conservatives regularly failed to fill all their allocated seats on the Science and Technology Committee and at the time of writing two of Labour's seats stand vacant. One has remained unfilled for more than a year.

In his famous 'Two Cultures' lecture of 1959, C. P. Snow posed a challenge: 'A good many times I have been present at gatherings of people who, by the standards of the traditional culture, are thought highly educated and who have with considerable gusto been expressing their incredulity at the illiteracy of scientists,' he said. 'Once or twice I have been provoked and have asked the company how many of them could describe the Second Law of Thermodynamics. The response was cold: it was also negative. Yet I was asking something which is the scientific equivalent of: *Have you read a work of Shakespeare's?*

'I now believe that if I had asked an even simpler question – such as, What do you mean by mass, or acceleration, which is the scientific equivalent of saying, *Can you read?* – not more than one in ten of the highly educated would have felt that I was speaking the same language. So the great edifice of modern physics goes up, and the majority of the cleverest people in the western world have about as much insight into it as their neolithic ancestors would have had.'

Were Snow's challenge to be repeated today in Parliament or Congress, it is hard to believe that more than one in ten members could rise to it. Few politicians have even a firm appreciation of the methods of inquiry that revealed the Second Law of Thermodynamics in the first place. That in large part reflects the backgrounds of the men and women who are elected to office.

There are 650 MPs in the House of Commons. Some 158 have a background in business and ninety were political advisers or organizers. There are eighty-six lawyers, and thirty-eight journalists and publishers. These professions bring valuable perspectives, but another one is all but missing. A solitary MP – Julian Huppert, the Liberal Democrat member for Cambridge – worked as a research scientist before beginning his political career. Only two of his colleagues even have science PhDs: Stella Creasy, the Labour MP for Walthamstow, in psychology, and Therese Coffey, the Conservative MP for Suffolk Coastal, in chemistry.

The wider scientific credentials of the House of Commons are decidedly weak too. An analysis by *The Times* after the 2010 general election identified only about seventy MPs who had shown any consistent interest in the subject at all. The higher up you go on the political food chain, the worse this representation of science gets.

Only one of the twenty-three members of the current Cabinet has a science background: Vince Cable, the Business Secretary, read natural sciences at university (though he switched to economics). In the previous Labour Cabinet, John Denham and Margaret Beckett were the only science graduates.

The situation is no better in the US, where the current House of Representatives includes among its 435 members just one physicist, one chemist, one microbiologist and six engineers. Another sixteen congressmen are medical doctors, and there are two psychologists, two dentists, a veterinarian and an ophthalmologist. There are no research scientists or engineers at all in the Senate, and just two medical doctors, a vet and an ophthalmologist. Lawyers make up 38 per cent of the House and 55 per cent of the Senate.

There are, of course, some politicians without academic training in science, medicine and engineering who have a geek's affinity for what they have to offer. David Willetts, the UK's current Science Minister, a philosophy, politics and economics graduate, is a good example, as is Phil Willis, a former history teacher who became an effective chairman of the Commons Science and Technology Committee. Henry Waxman, the California congressman, a lawyer and political science graduate, is another politician who effectively champions science and its methods.

The more normal attitude of politicians drawn from law, business and the humanities, however, is studied indifference. Before the last election, Adam Afriyie, then the Tory Shadow Science Minister, pledged to hold compulsory science seminars for new Conservative MPs as part of their induction course. When the Parliamentary Office for Science and Technology held such a session, open to all parties, it had become optional and barely a dozen MPs turned up. The sparse attendance included people like Huppert, who least need insight into how science can help them to do their jobs. Even MPs who profess an interest sometimes do so out of lip service. A recent participant in the Royal Society's pairing scheme, which links scientists and politicians, repeatedly forgot his pair's name and described him as 'the work experience'.

David Cameron took eighteen months after becoming prime

minister to make a major speech about science and technology. When three UK scientists won Nobel Prizes in 2010, he omitted to congratulate them. Tony Blair extolled science in a speech to the Royal Society in 2002, but he had been five years in Downing Street before he got around to making it. It's hard to imagine such a long silence about health, or education, or defence. Science just isn't a priority for the majority of the political classes.

President Obama, who pledged to 'restore science to its rightful place' at his inauguration and spoke of America's need for a 'Sputnik Moment' of scientific inspiration in the 2010 State of the Union address, stands out as an exception. But his efforts to promote science in Washington have been stymied by an unsympathetic Congress. His proposals for significant increases in science spending have been blocked, and as this book went to press, the budget compromise agreed in summer 2011 seemed certain to cut the funds available for research.

Even Obama has made little of science on the campaign trail. In 2008, indeed, both he and his main opponent in the Democratic primaries, Hillary Clinton, refused an invitation to debate science at the Franklin Institute in Philadelphia, after 38,000 people signed the ScienceDebate2008 petition. John McCain, the Republican front-runner, didn't even bother to reply.

Two cultures

Why is it, then, that so few geeks – whether they be scientists or non-scientists with a passion for science – make it far in front-line politics? One explanation lies in precedent: there simply aren't many examples of scientists serving successfully in high office to emulate, at least in the US and UK.

Stephen Chu, the Nobel physics laureate and US Energy Secretary, is appointed, not elected. It's hard to think of a contemporary English-speaking equivalent of Angela Merkel, a research chemist who became German Chancellor: a political generation has passed since Margaret Thatcher, another chemist, left 10 Downing Street. China's senior leadership (though unelected) is dominated by engineers and scientists – President Hu Jintao is a hydraulic engineer, and Premier Wen Jiabao is a geologist – to a degree that's unthinkable in Britain or America.

Then there is the question of opportunity. The structure of a scientific education and career doesn't really lend itself to active involvement in politics. In an era in which politics is becoming increasingly professionalized, it is hard for scientists, who must establish themselves with long, hard hours in the laboratory, to acquire the contacts and campaigning experience they'd need to make the switch.

This starts early. Student politics, in which many aspiring leaders cut their teeth and begin networking, tends to be the preserve of those reading humanities and social sciences – partly because they simply have more time. The demands of lab and lecture theatre mean that students reading medicine, engineering or physics typically have twice as many scheduled teaching hours as their peers studying history or languages. As the University of Bristol tells prospective arts students: 'If you have friends studying science or engineering, they will be either horrified or deeply envious (or both) that you may have only six or eight hours of classes per week.'

Most of all, science and politics have fundamentally different cultures that can make it hard to cross from one domain to the other. In science, every idea must be tested against evidence: it follows what Carl Sagan called 'the hard but just rule ... that if the ideas don't work, you must throw them away'. These exacting standards come with a kinder side: science is forgiving, in that perhaps more than any other pursuit it is comfortable with being wrong. There is no shame in having a bad idea that fails to survive contact with the evidence, so long as it is not clung to in defiance of the data. Scientists are allowed – encouraged even – to change their minds.

Richard Dawkins is fond of recounting an anecdote from his undergraduate days that illustrates this beautifully: 'There was an elderly professor in my department who had been passionately keen on a particular theory for, oh, a number of years, and one day an American visiting researcher came and he completely and utterly disproved our old man's hypothesis. The old man strode to the front, shook his hand and said, "My dear fellow, I wish to thank you, I have been wrong these fifteen years." And we all clapped our hands raw. That was the scientific ideal, of somebody who had a lot invested, a lifetime almost invested in a theory, and he was rejoicing that he had been shown wrong and that scientific truth had been advanced.'

This facility with the U-turn is bred into researchers from the moment they set foot in a laboratory. If it can sometimes be difficult to achieve in practice, every professional scientist can remember a clever hypothesis they've had to jettison when it became clear they were wasting neurons. It is an attitude, however, that rapidly has to be unlearned should a scientist wish to make progress in a different realm of public life. What science admires as intellectual honesty is seen in Westminster and Washington as the stuff of the gaffe.

What science admires as intellectual honesty is seen in Westminster and Washington as the stuff of the gaffe

For evidence, look no further than Senator John Kerry. On war in Iraq, the death penalty for terrorists, affirmative action and President Bush's 'No Child Left Behind' education reforms, Kerry's propensity for revising his position saw him attacked as a serial flip-flopper when he ran against Bush in the 2004 election. Never mind that some of his rethinks were subtle shifts in response to new information – many people changed their minds about war in Iraq when it emerged that there were no weapons of mass destruction, and he withdrew his support for Bush's education reforms when they were inadequately funded. His U-turns saw him labelled as a man whose opinion you couldn't trust, most memorably in the Bush campaign ad that featured a windsurfing senator with the slogan: 'John Kerry: whichever way the wind blows'. He will be forever haunted by his remark about funding for the Iraq war: 'I actually did vote for the $87 billion before I voted against it.'

Successful politicians prefer to proclaim the unshakeable strength of their convictions. But while this might sometimes help voters to decide whether a candidate or party can be relied on to behave a certain way, it can be anathema to evidence-based policy. Tim Harford, the *Financial Times* columnist and author of *The Undercover Economist*, explains the drawbacks eloquently. 'Tony Blair boasted of having no reverse gear,' he says. 'Margaret Thatcher said U-turn if you want to, the lady's not for turning. That's all very well, but would you drive a car with no reverse gear, or that wouldn't U-turn?'

Bill Foster, a Fermilab particle physicist who served as a US congressman for Illinois from 2008 to 2010, says the different demands and disciplines of politics can be infuriating to geeks. 'It's a nasty job, politics. The fundraising and campaigning are constant. You have to spend a large amount of your time doing non-thoughtful things, and there's the necessity of dumbing down your message. On the campaign trail, you're lucky if voters hear you say anything at all. You have to keep repeating your message, so that if they are listening that's what they'll hear. As a scientist, it makes you feel like an idiot, but it has to be done. It's tough. Working hard comes naturally to scientists, and so does mental discipline, but simplifying doesn't.'

What we're left with is a political class with too few geeks, and hence without an instinctive appreciation of what science is, how it works, what it does, and what it needs to deliver maximum benefit to society. And this has a number of pernicious consequences.

The wrong thinking

Richard Feynman famously described science as 'what we have learnt about how to keep from fooling ourselves'. It is easy to get carried away with a beautiful idea, whether it concerns social justice or the structure of the atomic nucleus. When we want to believe in something, all of us are tempted to note only the evidence that supports it, and to avoid or dismiss any that points another way. Psychologists call this confirmation bias, and it is a powerful source of error.

We also like to think of ourselves as clever, thoughtful, rational and right, and facts that suggest we are mistaken can pose a powerful challenge to that self-image. They induce what is known as cognitive dissonance: the uncomfortable feeling that emerges from holding contradictory beliefs at the same time. To reduce that dissonance, we are as likely to reject the evidence, or to explain it away, as to change our minds. This is especially true when accepting our errors might make us look weak or stupid.

Geeks and scientists are as susceptible as anyone else to confirmation bias and cognitive dissonance, but we have a great advantage. We know about these risks, and can thus seek to defend against them. The methods of science, indeed, have been devised as bulwarks against these powerful forces.

Their essence is to look for countervailing evidence, to test a hypothesis before coming to the conclusion that it is provisionally correct. People who do science are trained in these skills, the value of which is also clear to those who read widely in it. 'Remembering to attend to counter-evidence isn't difficult, it is simply a habit of mind,' wrote Kathryn Schulz in *Being Wrong*, a book which every politician should read. 'But, like all habits of mind, it requires conscious cultivation. Without that, the first evidence we encounter will remain the last word on the truth.'

In politics, this habit of mind is far too rare. Its scarcity contributes to a strained relationship between politicians and scientific advice that causes policy failure again and again. As we'll see in the next chapter, politicians are keen to give their policies the imprimatur of being evidence-based, without committing to the rigorous standards that requires. 'They want the authority of the white coat,' says Alan Leshner, chief executive of the American Association for the Advancement of Science and a former US government adviser, 'but without the obligations that confer that authority.'

This poor grounding in science isn't limited to its methods. Far too many candidates for public office also show staggering ignorance of its conclusions, to the point of rejecting robust scientific facts. Just look at the field in the 2011–12 Republican presidential primaries.

Michele Bachmann regards climate change as a hoax based on 'manufactured science'. Rick Perry holds similar views, describing global warming in his book, *Fed Up!*, as a 'contrived phoney mess that is falling apart'. Both candidates also dismiss the fact of evolution: they are supporters of 'intelligent design' who want it taught in schools. Bachmann has even invented fictional Nobel laureates who, she claims, agree with her. To curry favour with primary voters, Mitt Romney has backtracked from previous statements that climate change is happening and humans are responsible, and Jon Huntsman is seen as a rank outsider because he accepts science. 'To be clear,' Huntsman tweeted, 'I believe in evolution and trust scientists on global warming. Call me crazy.'

It isn't just the Tea Party Republicans. Politicians of all parties have been keen to promote scares that have been investigated and dismissed – such as the refuted link between vaccines and autism – or that fairly simple scientific principles reveal as improbable. The alleged cancer risks from mobile-phone masts are a good example. Mobile phones use non-ionizing radiation, which by definition cannot break chemical bonds and damage DNA to cause cancer. The inverse square law also means that exposures reduce very rapidly with distance. If a handset won't give you a tumour, and there's no evidence it will, a mast certainly won't. Yet dozens of British politicians have highlighted these phoney risks in the House of Commons. Three past or present MPs are even trustees of a campaign group.

When Bill Foster was elected to Congress in 2008, he was amazed at the levels of ignorance about science, and the disdain for it, that he encountered on both sides of the aisle. He remembers a Democratic colleague who argued that 'we should really put more windmills into our energy policy because they poll well', without any regard for whether they would actually address America's energy needs. A Republican, meanwhile, praised 'all the wonderful things the free enterprise system can achieve when government gets out of the way – things like planting crops using GPS'. The congressman had no idea that the US government had put up the funds to send the satellites that make up the global positioning system into orbit.

No MP or congressman needs to be expert in any particular aspect of science. The narrow but deep knowledge of the true specialist is rarely required by prime ministers and presidents. But it is deeply concerning that so many politicians with realistic aspirations to office are unable to tell science from pseudoscience. It is impossible to make sensible policy about climate change, medicine or food safety without a modicum of understanding of relevant facts.

Our leaders need to be intelligent consumers of science. They must know how to interrogate purportedly scientific findings to judge their reliability. They should be able to recognize rules of thumb by which scientists evaluate others' work, such as good controls, peer-reviewed publication and replication, and should have a feel for spotting extraordinary claims that require extraordinary evidence. They should also understand how these scientific tools can be exploited to evaluate policy.

The backgrounds from which the political classes are mostly drawn mean that these skills are too often missing. And a lack of familiarity with science also contributes to a failure of leadership. Without an instinctive appreciation for how science works and the circumstances in which it thrives, politicians let it down time and again.

The price

In 2001, President Bush thought he had designed an elegant compromise between supporters and opponents of embryonic stem cell research. There would be a ban on federal funding of studies that used newly created 'lines' of embryonic stem cells, but grants could still be awarded for studying cell colonies that were already in existence. It satisfied neither side, while betraying profound ignorance of what this promising field of medical research actually involves.

Very few of the sixty-odd lines approved by Bush were close to being of good enough quality for scientific research. All of them had been created using mouse cells to promote their growth, which would make them unsuitable for therapeutic use. The different rules governing funds from federal and other sources created a Byzantine nightmare for scientists who were in receipt of both. Laboratories had to buy everything in duplicate, from microscopes to paperclips. Kevin Eggan's lab at Harvard introduced a sticker system – green for kit that was privately funded, red for everything else. 'I've spent the last three years of my life trying to get this sorted,' Eggan told the *Guardian* in 2006. 'At least a third of my time is still spent keeping the accounts and equipment separate.'

It was a compromise that could only have been designed by politicians and administrators without the slightest inkling of how scientists actually do their science, and it was far from unique. Regulations, funding decisions and other policy initiatives are routinely introduced in Britain as well as America with scant regard for their effects.

Before the British general election of 2010, the Conservatives pledged to bring in a cap on immigration from outside the European Economic Area if elected – a promise that survived the party's coalition agreement with the Liberal Democrats. An interim cap was announced, and it soon became clear that it was having a devastating impact on British science. Research is an increasingly international pursuit, in which the leading nations compete to attract the best talent from all over the world, at every career stage from postdoctoral researchers to professors. Academic institutions, however, were now being given strict quotas on the number of visas they could issue, which severely limited their capacity to recruit.

The Institute of Cancer Research in London, which employs scientists from fifty-five countries and is Britain's top-rated academic institution, was told its visa allocation was being cut from thirty to four. The Beatson Institute for Cancer Research in Glasgow needed six visas and was originally offered just one. The quota issued to fifteen top research universities was cut by 233, a reduction of 21 per cent. Eight Nobel laureates wrote to *The Times* to protest that the measure was damaging British research. They included Andre Geim and Kostya Novoselov of the University of Manchester, newly minted winners of the 2010 Nobel Prize for Physics; both were born in Russia and might never have made it to Britain with a cap in force.

The government didn't mean to damage Britain's international competitiveness in science like this, and the cap's effects on research were eventually watered down following lobbying from the Campaign for Science and Engineering and reports of its damaging impact in *The Times*. That such a measure was ever even tabled, though, reflects science's position as a political afterthought. Senior politicians with a background in science, or who keep properly abreast of it, would not have needed to be told how detrimental the cap would be.

Political ignorance of the practice of science has also contributed to the introduction of medical research legislation with unintended and onerous consequences. In 1999, a public inquiry revealed that doctors at Alder Hey Children's Hospital in Liverpool had removed and stored organs from more than 850 deceased infants for research, often without obtaining informed parental consent. That was already illegal, and Dick van Velzen, the pathologist primarily responsible, was later struck off by the GMC, but the scandal led to demands for much tighter regulation of medical research. The result was the 2006 Human Tissue Act, which introduced tough new requirements for explicit consent for any tissue besides hair and nails used in research.

Regulations like these sound benign, and they were certainly well meant. But explicit consent is expensive and time-consuming to collect, and insisting on it when it is not strictly necessary to protect patients adds significantly to the costs of research that ultimately improves healthcare. The study of tumour tissue removed in biopsies, for example, is important to understanding how genetic mutations drive cancers and affect the way they respond to drugs. The need to ask patients for consent, instead of allowing them to opt out if they have strong views, makes this work more expensive. A 2008 study found that the Act's requirements cost one university £51,000 a year. 'Complying with the regulations costs money that would otherwise be spent on research,' Peter Furness, vice-president of the Royal College of Pathologists, told the *Guardian*. 'It also inhibits research, especially small-scale research . . . it is phenomenally difficult to organize because of the bureaucracy.'

The Human Tissue Act isn't the only source of red tape that has constricted medical research in Britain of late. The 2004 European

Clinical Trials Directive regulates all trials according to the tight controls needed only for the most sensitive ones. Sir Michael Rawlins, who led an Academy of Medical Sciences review of regulation published in 2011, described it as a disaster. The Academy's report revealed how researchers must often seek similar permission from a dozen or so different bodies before embarking on research involving patients as subjects, when one or two would suffice.

These restrictions mean that once the charity Cancer Research UK has funded a trial, it takes an average of 621 days before the first patient is treated. If scientists want to test an old drug for a new purpose, they must get the same lengthy ethical authorization needed for a completely new medicine, even if its safety profile is well known. Another European initiative, the Physical Agents Directive, would have banned much medical use of magnetic resonance imaging scans had not scientists eventually lobbied the European Commission to see sense.

'The regulation only ever expands,' says Sir Mark Walport, director of the Wellcome Trust, Britain's biggest biomedical research charity. 'The bureaucracy has become so onerous, without doing anything to deliver safer trials or better drugs.'

These are problems that could easily have been avoided if politicians and civil servants had a little more awareness of how regulations might affect science in practice. Science must of course respect the ethical boundaries that society quite properly sets. But those boundaries must also be set with respect for the realities of research if they are not to stifle it.

Poorly framed and unwieldy regulations have also added significantly to the bureaucracy and costs involved in animal experimentation, often without delivering improvements in animal welfare. The 2011 Bateson Report into primate research noted that the costs of such experiments increased fourfold between 2003 and 2008 because of measures such as standardizing humidity that can actually worsen the living conditions of monkeys.

Of still greater concern, the political system let down the researchers who conduct this important and perfectly legal area of biomedical science for many years by failing to protect them against the violent tactics of animal rights extremists.

In the late 1990s and early 2000s, a small group of British anti-vivisectionists waged a vicious campaign of intimidation against scientists who experiment on animals. Professor Colin Blakemore, an Oxford University neuroscientist who was a vocal defender of such work, was attacked and letter bombs were addressed to his children. Employees of Huntingdon Life Sciences, a testing company, were visited at home by activists, who vandalized their cars and daubed threatening graffiti on their houses. Its chief executive was assaulted with a baseball bat. A Cambridge University neuroscience laboratory was cancelled following police concerns about potentially violent protests. College boathouses and cricket pavilions were firebombed when Oxford University planned a bio-medical laboratory with animal facilities, and frightened builders dropped out.

Ministers made the usual statements about deploring violence without doing anything substantial to stop it. Some of their actions, indeed, gave succour to the extremists. In 2000, the governing Labour Party caved in to activist pressure and cravenly sold its pension fund's shares in Huntingdon Life Sciences. In 2003, leaked documents showed that Blakemore had been blackballed for a knighthood because of his 'controversial work on vivisection'.

It took geek activism to turn this tide. In 2002, Laura Cowell, a sixteen-year-old girl with cystic fibrosis, agreed to front a counter-campaign extolling the medical benefits of animal experimentation, and her bravery gave many more scientists the confidence to stand up for themselves. The Science Media Centre, launched in 2002, helped them to engage with journalists and the Association of the British Pharmaceutical Industry began warning ministers that valuable businesses would soon leave the UK unless they got a grip. Another teenager, Laurie Pycroft, formed a group called Pro-Test to support the Oxford lab, which staged a march of more than 1,000 people through the city in 2006. 'I'm sick of seeing only the anti-vivs' argument being represented,' he told the rally. 'It's time to speak out in support of scientific research.'

All this effort eventually paid off. In 2005, the government created new laws against harassment and intimidation. As Pro-Test built the pressure, the police's National Extremism Tactical Co-ordination Unit began to use its new powers against animal rights terrorists. In 2007, many of the ringleaders of the Huntington Life Sciences campaign were jailed. Oxford's biomedical research centre was completed with government support and opened in 2008. As more and more scientists took time to explain the case for animal experimentation, public support increased: the proportion not concerned about vivisection, or who are prepared to accept it when there are no alternatives, almost doubled from 32 per cent in 1999 to 60 per cent in 2009.

Making votes count

What Britain's experience with animal experiments shows is that we get, to a point, the politicians we deserve. As geeks stood up for an issue we cared about, the political inertia that allowed violent protesters to flourish was gradually overcome. If we want to raise the level of scientific understanding in politics, we have to repeat this time and again. We have to make such ignorance and indifference much harder to get away with.

To do that, we're going to have to mobilize. If we want to change politics for the better, the first thing we have to change is our own political behaviour.

As things stand, politicians aren't really accountable for decisions that affect science in the same way that they are over other issues, such as taxation, foreign policy or health. That's because science has never been able to command much of the currency in which that accountability is measured: votes. Even for most geeks, science is well down the list when it comes to deciding how to vote. This low political profile leads to a democratic deficit: as governments and MPs can be elected without significant scrutiny of their plans for and familiarity with science, they do not have to heed the views of those who care about it.

Candidates like David Tredinnick are able to be selected for safe seats and then to win them time after time, not because their constituents particularly value their dismissive approach to science, but because not enough of their constituents object strongly enough to change their votes.

Equally, vocal advocates of science have often lost, even in places where the 'science vote' ought to be strong, because their positions don't translate into support. Evan Harris, the last Parliament's most vociferous supporter of science, lost the Oxford West and Abingdon constituency in 2010 by just 176 votes, despite a track record that ought to have appealed strongly to Oxford University scientists and students. In the same year, Bill Foster lost the Illinois 14th District, which includes Fermilab – the particle physics laboratory where he worked.

If we want to elect and retain political champions like these, we're going to have to start thinking more creatively about how we use our votes. Of course, geeks have strong opinions about all sorts of issues other than science. Like everyone else, we have views about economics, social justice, foreign policy and taxation, and these will generally form our political standpoints and guide our voting choices at least as strongly as our views about science. That doesn't mean, though, that we can't begin to geek the vote, and get political parties and candidates to start taking science more seriously.

During the 1997 election campaign, the editor of *The Times*, Peter Stothard, took a somewhat unusual approach to the paper's endorsement. He decided that the mood of the next Parliament on one particular issue was more important than the identity of the winning party. That issue was Europe. His leading article encouraged support for Eurosceptic candidates, regardless of the political party to which they were affiliated, and the paper compiled a list for those voters who wanted to take its advice.

This unorthodox strategy had little effect, buried as it was in a New Labour landslide. But elements of it are worth reviving to promote more scientifically literate politics. In some constituencies – chiefly those with universities or where science-based industries are big employers – the number of scientifically motivated voters will easily outweigh the majorities of the sitting MPs. The majority in Oxford West and Abingdon is 176. In Cambridge, it is 6,792. There are further suitable seats in York, Manchester, Edinburgh, Norwich and London. Guildford is attractive: it is home to a cluster of high-tech companies around the University of Surrey, and the MP's majority is 7,782, up from just 347 in 2005. The Conservative incumbent, Anne Milton, is also a Health Minister who has defended NHS provision of homeopathy.

A few well-organized campaigns supporting candidates with strong scientific credentials, or against those who have let it down, might easily be capable of deciding a contest or two. A couple of surprising results, and the parties would quickly take notice. MPs elected this way will want to thank their supporters by standing up for science. Even those who survive against strong geek opposition will have an incentive to take their interests more seriously lest they be targeted again.

It's interesting to note that the Conservative candidate who defeated Harris in Oxford, Nicola Blackwood, has made assiduous efforts to court science and scientists since her narrow victory. That can only be a good thing. The more MPs who inform themselves about science and give geeks a sympathetic hearing the better, and political self-interest is the strongest lever we have got.

There's another approach to geeking the vote that could have particular utility in the British political system, where aspiring politicians often fight a no-hope seat before being selected for one where they have a decent chance of winning. With the first-past-the-post electoral system confirmed by the 2011 referendum, many contests in safe seats are foregone conclusions. But this opens a new approach to tactical voting.

If a candidate has a background in science, or a strong record of supporting it, she may well be worth voting for even if she has no realistic chance of victory. If a Tory scores 1,000 votes better than expected in a losing cause in Barnsley, she will stand a much better chance of being selected to fight Woking or Basingstoke next time. For the many people living in safe seats who consider their votes wasted, this is a way of making them count. It won't materially affect the result of the present election, but it could help to build up a cadre of serious figures in every party who understand science, who can influence policy when their time comes.

This geeky take on tactical voting needs easy access to data about candidates' views and positions if it's to stand any chance of success. Fortunately, that sort of information has never been more readily available: the internet has made it cheap and easy to collect. Just before the 2010 general election, James O'Malley and Craig Lucas set up a website called Skeptical-Voter.org, a wiki that collects details of candidates' views on a laundry list of scientific and skeptical issues. It was far from comprehensive, it started too late to have much of an impact, and it didn't get enough publicity. But with three years to go until the likely date of the next election, there is more than enough time to turn this into a terrific database.

Websites such as TheyWorkForYou.com have made it simple to find out any sitting MP's voting record on particular issues of concern. With online social networking tools such as Twitter and Facebook growing in reach, this information is easily added to and shared. Time was when, to find out candidates' positions on science, you'd have had to go along to hustings and hope you were called to ask a question. Not any more.

The political parties' records and manifestos can also be subjected to geek audit. During the 2010 election campaign, Martin Robbins, who writes the *Lay Scientist* blog for the *Guardian*, sent detailed questions to each of the main parties, then assessed their answers and summarized their policy positions for people who wanted to 'vote science'. The Campaign for Science and Engineering also solicited and published letters from the party leaders about their plans for science.

If you're going to support a party or candidate because of science issues, don't keep quiet about it: we need them to know why they're getting your vote. Write to the candidate or her campaign. Blog about your decision. Explain your position to party activists who canvass you. It's important to spread the message among politicians at every level that science is an issue that motivates people at the ballot box. They need to know that geeks vote too.

We need to be careful, though, that in making science a political issue we don't allow it to become a polarizing one. It would be dangerous for its interests to become too closely aligned with those of a single party. The risk is that the other side will see us as hostile opponents they will never win over, and fail to give us the hearing we deserve. There is even the possibility that that party will seek to punish science when it comes to power.

Daniel Sarewitz, Professor of Science and Society at Arizona State University, who writes a political column for the journal *Nature*, has argued that excess politicization of this sort has been a significant factor in the damaging breakdown in relations between science and the Republican Party in the US. The close identification between most scientists and the Democrats may even have encouraged the Bush Administration and House Republicans to take more extreme positions on issues such as climate change and stem cells. Attacking science became a way of pandering to the base and baiting people who weren't seen as 'one of us'. That kind of political narrative helps nobody.

This doesn't mean that, when a particular party systematically engages in science abuse, anybody should take it lying down. Many Republicans have come to take positions that can fairly be characterized as anti-science, as have many Greens, and these need to be countered robustly. It's important, though, to cultivate allies in as many parties and caucuses as is possible. Our powers of influence are greater when we can count on people inside the tent.

In that spirit, *The Geek Manifesto* doesn't contend that there's a mainstream party that is particularly deserving of our votes, or one that we should never countenance supporting. Most have candidates who are both good and bad on scientific issues. The Liberal Democrats, for instance, are the party both of Evan Harris and Julian Huppert, and of Norman Baker, who opposes animal experiments and promotes conspiracy theories about the death of David Kelly.

The dirty business

David Tredinnick was always going to retain his seat at the 2010 election. Parliament's arch-advocate of alternative medicine and astrology held on to the Bosworth constituency even during the Labour landslide of 1997, and he was defending a majority of more than 5,000. So when Michael Brooks decided to stand against him, he knew he wasn't about to win. Brooks, a science writer, had a different goal: he wanted to make sure that the voters of Bosworth understood precisely who they had been electing as their MP year after year.

Describing Tredinnick as 'a champion of pseudo-science and a hindrance to rational governance', Brooks formed a one-man Science Party. Simon Perry, who runs Skeptics in the Pub in nearby Leicester, helped him to find ten Bosworth residents to sign his nomination papers and he handed over his £500 deposit. He'd never get it back – in fact he lost it in spectacular fashion, clocking up just 197 votes against the MP's 23,132. But it bought him places at hustings, airtime on local radio and space in local newspapers to draw attention to his opponent's strange views.

Tredinnick's majority narrowed a little amid a national swing to the Conservatives, though it's impossible to say whether Brooks was responsible. More importantly, he gave the Tory voters of Bosworth no excuse for being unaware of what they were supporting. He made it impossible for the candidate's embarrassing opinions to be swept under the carpet and forced him to defend them in public.

Such stunt campaigns will probably have an occasional role to play in the geek movement, by turning the positions of particularly unfriendly politicians into local issues. Brooks, though, is the first to admit that his Science Party has no broader future. There's no opening for a single-issue party like the Greens to promote science. Such a strategy could actually be counterproductive, by presenting science as a special interest rather than a subject with which every party must properly engage.

What's really needed is for every party to include more geeks among their elected representatives. And that will mean more of us getting actively involved in party politics and ultimately running for office. 'There's a cultural problem here, in that scientists often see getting involved in politics, getting involved in government in general in fact, as pretty dirty,' says Bill Foster. 'But there's no substitute for getting technically trained people into Congress. What you get from scientists and engineers is an instinctive thought process that is hugely valuable. The evidence-based approach comes naturally, as does a more rational attitude to risk – something that's frankly missing in our society.'

John Denham, one of the few recent Cabinet ministers with a science degree, agrees. 'It's always a good idea to have people with a relevant background,' he says. 'More scientists and engineers in political and public life would be better, certainly from this starting-point. They bring a knowledge and insight into the activities of science that's in short supply.'

Add a few more scientists to the mix at Westminster and Washington, and you'd rapidly raise the game of the non-scientists who surround them. 'If you could get maybe 25 per cent with a background in science and technology, their expertise would diffuse into the whole culture of Congress,' Foster says. They would bring a different perspective both to caucus meetings and to more casual settings, which can be just as important. It is often the informal discussions between colleagues in the bar, the lobby and the tea room that do most to shape the political mood; with so few scientists in Parliament and Congress, they simply can't participate. Not only does the input they'd provide go unheard, but the issues they'd raise don't even make it into conversation.

Reversing this may well prove difficult. As we've seen, there are many reasons why scientists are often reluctant to immerse themselves in active politics. If we can make science a more salient political issue, however, that could soon start to change. The existence of a recognizable 'geek vote' would start to turn candidates with a background in science into more tempting choices for the parties, and their chances of electoral success would improve too.

Before such candidates can be selected, we need to make sure there are enough of them in the pipeline. Science is poorly represented at every level in politics – among the councillors, state representatives and activists from whom prospective parliamentary candidates and primary contestants are chosen in the first place, as well as in the higher echelons. We need more geeks to get involved lower down the food chain if we're eventually to see more of them among the big beasts. 'I went straight from science to Congress, but that's atypical,' says Foster. 'You want people with a record, but few scientists have even stood for their school board.'

Nobody is going to join a party, let alone stand for a council election, unless they have strong political motivations that run beyond science. Plenty of us, however, already have an allegiance to a party, and at least some of us are going to have to get more involved. If you're a scientist and a Tory (or anything else), join the party. You may get a chance to influence policy, and you'll certainly have a voice in selecting candidates. The constituency party meetings that choose their parliamentary candidates are often attended by a hundred or so members and can be settled by majorities of little more than a dozen. There's ample opportunity for a few newcomers with science in mind to make a real difference.

'There's really no alternative to engaging in electoral politics,' says Denham. 'People in this movement need to break out of the habit of scientists only talking to other scientists.'

The parties may be more receptive to this than you might think. 'The intake of politicians is becoming narrower for all parties,' says Denham. In the aftermath of the expenses scandal, they are desperate for new blood, and for candidates with real experience in diverse walks of life. Some of that could be geek blood. It's a good time to make a mark.

Bill Foster is thinking radically about how best to achieve this. In 2011, he joined another physicist who made it to Congress, the Republican Vern Ehlers, to launch Ben Franklin's List. This bipartisan political action committee is explicitly devoted to getting more scientists into politics by providing their campaigns with the financial assistance they need to succeed. Named after one of the

Founding Fathers of American democracy, who doubled as an accomplished physicist and inventor of the lightning conductor, the initiative is modelled on Emily's List, the highly successful initiative to get more women elected to office. From a standing start in 1985, eighty-four congresswomen, sixteen senators and 584 state legislators have now been elected with support from Emily's List. If Ben Franklin's List can even begin to emulate those achievements, the scientific complexion of Congress would be transformed.

'It's going to be tough sledding,' says Foster. 'Part of the purpose is to fill the pipeline. If we can get a lot of people into state legislatures over the next few years, we can create a better pool of suitable candidates.'

There are risks to this approach. Initiatives such as Ben Franklin's List can't be seen to be too partisan, and they mustn't be allowed to spread an erroneous message that only trained scientists deserve the geek vote. 'An Emily's List for science could have a very negative effect on the likes of Phil Willis, non-scientists who are good on science,' says James Wilsdon, Professor of Science and Democracy at the University of Sussex, who was until last year director of policy at the Royal Society. 'We must take care not to say that only scientists are qualified to speak about and for science.' Nevertheless, carefully framed, initiatives like this – whether formal or informal – could have a beneficial impact. The money is almost certainly there: there's no shortage of high-tech business leaders and entrepreneurs with the resources to contribute.

Whether we try to finance the election of more geeks or not, Wilsdon is right that we shouldn't ignore the non-scientists who are always likely to make up the majority of the political classes. One of the virtues of electing more scientists, indeed, would be to improve by osmosis the scientific awareness of the other MPs, congressmen and senators around them. And while our voting behaviour is an essential means of holding these people to account and making them take science more seriously, it isn't the only one. The need to engage actively with politicians doesn't stop when the ballots have been counted. That's when the governing begins for the winners and the lobbying starts for everyone else.

The redress of grievances

Sean Tipton, director of public affairs at the American Society for Reproductive Medicine, often advises doctors and patient groups about how best to make themselves heard in Washington. He always asks them to recite the First Amendment to the US Constitution.

'They always know that it guarantees freedom of religion, speech and assembly,' he says. 'I have to remind them that it also guarantees the right "to petition the Government for a redress of grievances".'

It's a reminder that political science should heed. While other groups with an agenda, such as anti-abortion and animal rights campaigners, are always in the ear of politicians, we often seem to forget that MPs, congressmen and senators are representatives, who are supposed to listen with an open mind to what their constituents have to say. We fail to make our case to politicians directly, and we even fail to do so through people whose job it is to lobby on our behalf.

Jon Spiers, head of public affairs at Cancer Research UK, says his team regularly meets with bemusement when it consults researchers funded by the charity about their policy concerns. 'We go out and tell them we're employed to lobby the Government, what can we do for you, and frankly we sometimes struggle. There's a sense of: "Oh, is that even possible?" Scientists tend to feel that politics is something that happens to them, not something they can influence.'

Such lobbying, particularly when it is well timed and well framed, can deliver dramatic results. When the British government published proposals to reform fertility and embryology legislation at the end of 2006, it was immediately clear to researchers in the field that its plans would cause them major problems. Of greatest concern was its intention to ban the creation of embryos that merge human DNA with empty animal eggs, which several scientists were already pursuing as a potential source of stem cells for researching devastating conditions such as motor neuron disease.

Their response was not just to complain, though they did use the media rather effectively to highlight both their concerns and those of the patients their work was ultimately intended to help. Scientists working in stem cell research, such as Stephen Minger, Alison Murdoch and Robin Lovell-Badge, also began to meet ministers, officials and MPs to explain precisely what they proposed to do and why they wanted to do it. The Department of Health soon back-tracked, and the government became an enthusiastic advocate for the research: Gordon Brown, the Prime Minister, wrote a supportive opinion piece for the *Observer* immediately before the free vote of the House of Commons that eventually approved it. Constructive engagement had successfully redirected the policy down a more favourable track.

An important factor in this campaign's success was its timing. It started well before the ministers responsible for the Human Fertilisation and Embryology Bill had made any irrevocable decisions or tied themselves too closely to a firm policy, and before most MPs had had to express an opinion in a vote. Starting early kept cognitive dissonance at bay. Caroline Flint, the Health Minister in charge, had yet to commit to a ban, making it easier for her to change her mind.

Lobbying serves a further purpose that is particularly important to geeks because so few of us hold political office. It makes mood music, putting science in front of politicians who do not normally think about it much at all. If enough constituents share their views with their MPs, we can force them to recognize that scientific issues matter to a significant subset of the people they represent, and to work out where they actually stand on questions that would otherwise have passed them by.

Nicola Blackwood, the new Oxford West and Abingdon MP, says that an overflowing inbox has a particular impact when it concerns a matter with which she's not especially familiar. 'What grabs my attention as an MP, given the demands on my time, is when signifi-cant numbers of people contact me,' she says. 'It gets the issue on my radar.' Environmental and international development NGOs do this all the time; geeks do not. 'You need to be the campaign group that gives ministers the biggest headache,' said Tracey Brown, director of Sense About Science. 'That way, they can't ignore you.'

Most politicians, thankfully, aren't David Tredinnick. If they let science down, it's usually because they know little about it, not because they're actively hostile. We don't necessarily have to change their minds, we just have to persuade them to look a little deeper.

'There's a myth among politicians that science is all too technical, that you won't understand it if you're not a scientist,' says Blackwood, a musician who confesses to feeling that way herself in the past. 'But you don't have to understand every detail to understand the goals, why they matter, and how the funding streams and infrastructure are important to them. There's almost always something I don't understand, but you don't need a deep technical knowledge in the way I thought, and in the way most of my colleagues think you do.'

'You need to be the campaign group that gives ministers the biggest headache. That way, they can't ignore you' – Tracey Brown

Her changed perspective emerged from another form of engagement between science and politics that has great potential to build the level of understanding that's required. In the winter of 2010–11, she was among sixteen MPs and twelve civil servants who took part in the pairing scheme organized by the Royal Society. Professor Dave Wark, a physicist and a Fellow of the Royal Society, spent a week in Blackwood's office at Westminster and the MP paid a return visit to Wark at the Rutherford Appleton Laboratory.

It was an eye-opening experience for both parties. 'Most scientists feel sort of helpless before the political process, like leaves thrown in the wind,' Wark says. 'It's made me realize that politicians often want to learn from us, that we have to get involved and help them out. I'm the first Fellow of the Royal Society to take part in the pairing scheme: I'd always said I was too busy. But you can't be too busy for this. It's too important.'

Blackwood agrees. 'I can't see why you wouldn't do it,' she says. 'I'd felt intimidated by scientists, but you learn not to be afraid. It breaks down barriers to mutual benefit.' Too few of her parliamentary colleagues share her enthusiasm for pairing. It's an opportunity that the parties should be doing far more to promote. A party that expected its MPs to take part in the scheme, and to take it seriously, would have a good claim on the geek vote.

We need to be proactive constituents as well, to make sure that our MPs know who we are and what we care about. We must go to constituency surgeries and invite local MPs to labs, to explain the research that goes on in their backyards. Leaders of the scientific community can use their rights as constituents too. 'My MP is Justine Greening, who's a Treasury minister, and you know I don't think I've ever actually met her,' admits Mark Walport.

Mary Woolley, President of Research! America, a US advocacy group, describes the challenge as 'the Starbucks test'. 'When she speaks to groups of researchers, she asks them if they would recognize their member of Congress if they were waiting in the same line for coffee,' *New Scientist* magazine reported. 'Most say yes. Then she asks: would they recognize you? Most hands go down. Woolley's point is that members of Congress know the top lawyers in their districts, the leading small business owners, and so on. And they see them as key players in the constituency they've been elected to serve. In this Congress, more than ever, scientists will need to ensure that they are part of this picture, too.'

If we can achieve this sort of recognition, we'll deserve a better class of politicians with a better appreciation of science. We'll start to get them, too, as they begin to realize they have to engage properly with science to keep their geek constituents happy and that doing so might even win them a few votes. We'll help politicians to help science, to give it the support it needs. In time, we might also start to address a further failing of modern politics, to which we're turning next: its reluctance to exploit science to design policies that are fit for purpose.

POLICY-BASED EVIDENCE

Why science matters to government

THE EXPERT ADVISER, SAID THE MINISTER, HAD BEEN ABUNDANTLY clear. 'What he said was that closing down special needs schools and putting needy kids into mainstream education is a lousy idea,' said Hugh Abbott, the Secretary of State for Social Affairs and Citizenship.

Malcolm Tucker, the splenetic and foul-mouthed Downing Street press officer from *The Thick of It*, had other ideas. 'Yeah, but I've got an expert who will deny that.' Abbott protested: who was Tucker's expert? 'I have no idea, but I can get one by this afternoon. The thing is you have spoken to the wrong expert. You've got to ask the right expert. And you've got to know what an expert's going to advise you before he advises you. Hugh, whether you like this or not, you are going to have to promote this Bill, so what I'm going to do is, I'm going to get you another expert, yeah?'

Politicians of all parties are addicted to evidence. They know that it is fundamental to grown-up and rigorous debate, and that without it their arguments will sound empty and groundless if they are lucky, opportunistic and self-serving if they are not. They want to appear pragmatic – committed, as Tony Blair used to put it, to the notion that 'what matters is what works'. They thus portray themselves as champions of evidence-based policy, who always make decisions according to the very best expert advice.

This commitment, however, is skin-deep. For all their rhetoric about evidence-based policy, what most politicians really value is something rather different. They want evidence, for sure: they require it if they're going to make a halfway convincing case. But that evidence should support the policy they want to implement; they don't want inconvenient data that suggests their initiatives won't work.

Like Malcolm Tucker, what they want is not evidence-based policy, but policy-based evidence.

What politicians want is not evidence-based policy, but policy-based evidence

When Gordon Brown's long wait to succeed Tony Blair finally came to an end in 2007, the new Prime Minister decided that one way to set him apart from his long-term rival would be to reverse the government's recent decision to reclassify cannabis as a class C drug. This had been done only three years previously, on the recommendation of the Advisory Council on the Misuse of Drugs (ACMD), which considered that the evidence of harm was insufficient to warrant class B status and a maximum penalty for possession of five years in prison. The new occupant of No. 10 felt this had sent damaging signals that the drug was safe. His Home Secretary, Jacqui Smith, ordered the ACMD to reconsider.

The advisory panel agreed to take another look. But it refused to dance to the Prime Minister's tune. It found, as it had done in 2004, that the evidence supported the classification of cannabis in class C, and twenty of its twenty-three members voted against a change in the law. Sir Michael Rawlins, its chairman, said that reclassification 'is neither warranted, nor will it achieve its desired effect'.

This wasn't the policy-based evidence that Smith and Brown wanted. So they ignored it. They sought out other, more congenial experts whose research suggested a stronger link than had hitherto been appreciated between cannabis and psychosis. That the ACMD had already considered this research and found it insufficient to warrant class B status mattered not. It was the right kind of evidence for the policy the ministers had in mind, so it trumped the council's verdict. In May 2008, Smith announced that cannabis would indeed be upgraded. 'There is a compelling case to act now rather than risk the future health of young people,' she said.

Malcolm Tucker would have been proud of her.

He might also have expressed grudging admiration for a minister from the other side of the political fence. In June 2010, a month after his appointment as Health Secretary in the new Conservative–Liberal Democrat coalition government, Andrew Lansley published a reform bill that would radically restructure the way Britain's National Health Service works. Primary Care Trusts, responsible for commissioning health services, would be abolished and replaced with hundreds of much smaller consortiums of general practitioners. The NHS's mortality rates, Lansley said, were so much worse than those from comparable countries that there was a manifest need for change: Britain's death rate from heart attacks, for example, was double that in France. There was, he said, 'a range of evidence' that GP commissioning was the best way to improve matters.

In health policy, if you're going to fight your corner this way, it's wise to be pretty certain of your facts. Doctors and health-policy wonks tend to be rather good with evidence: more than most other professions, they value it, they know how to handle it and they bother to check references. John Appleby, the chief economist of the King's Fund, a health think-tank, began to look carefully at the mortality statistic Lansley was quoting as justification for his radical reforms and quickly noticed that it didn't quite tell the story that the Health Secretary claimed.

The government's assertion was correct, so long as you look at a single point of data. Figures from the Organisation for Economic Cooperation and Development showed that in 2006 there were nineteen deaths per 100,000 people from heart attacks in France, and forty-one per 100,000 in the UK. This use of a single point of data, however, was profoundly misleading. Death rates from heart attacks are still higher in Britain than in France, but they are also falling much, much more quickly. 'Not only has the UK had the largest fall in death rates from myocardial infarction [heart attack] between 1980 and 2006 of any European country, if trends over the past thirty years continue, it will have a lower death rate than France as soon as 2012,' Appleby wrote in the *British Medical Journal*.

The evidence that Lansley's GP commissioning plan was the best means of improving standards did not fare much better when Ben Goldacre, the *Guardian*'s 'Bad Science' columnist, put it under his microscope. As the precise form of this proposed structure had never been tried before, there was nothing directly relevant in the literature.

Goldacre was charitable, and searched instead for studies of GP fund-holding, a broadly similar policy that had been tried out in the 1990s. He found only four research papers in the academic literature: 'Kay in 2002 found it was introduced and then abolished without any evidence of its effects. In 2006 Greener and Mannion found a mix of good and bad but no evidence that it improved patient care. In 1995 Coulter found nothing but gaps in the evidence and no evidence of any improvement in efficiency, responsiveness, or quality. Petchley found there was insufficient data to make any judgement. Lansley says he is following the evidence. I see no evidence to follow here.'

Both Smith and Lansley asserted that their policies were evidence-based because that made them sound properly thought through. They didn't want to give the impression of pandering to the tabloids in Smith's case, or of meddling with a popular national institution in pursuit of an ideological agenda in Lansley's. Neither policy, however, was actually evidence-based at all. The ministers did not reach their positions through careful deliberation of all the available data before coming to the conclusion. Rather, they took a position, then added what David Halpern, a senior adviser to both Tony Blair and David Cameron, has called 'spray-on evidence'. The conclusion comes first, followed by a trawl for useful data that might be daubed on afterwards.

Cargo-cult Politics

Evidence is the foundation of science, the standard against which new ideas must be tested and old ones re-evaluated. It is the requirement to test every claim against observable data that makes the scientific method an incomparably reliable way of generating knowledge. But as geeks understand, science has rules about using evidence.

It isn't a kind of magic dust you can sprinkle on an idea to give it credibility: its power to reveal and predict emerges only when it is considered in the right way. You have to look at all the available evidence, not just the convenient bits. And the evidence is supposed to come first: you're meant to reach your conclusions after you've tested your hypothesis against reality. You can't decide on your position, like Smith and Lansley, and then cherry-pick the data to suit.

Science understands this temptation and draws its strength from the safeguards its methods raise against it. Control groups and statistical analysis are used to show that apparent causes and effects are real and not a matter of coincidence. Findings must be published openly, with full data and working, so that others can pick them apart. Research is accepted for publication only after it has been peer-reviewed, to guard against sloppy errors and misinterpretation of data, and it is not considered robust and well founded until it has been independently replicated. Through adherence to these standards, scientists acquire the habit of questioning the evidence for their beliefs and taking note of contradictory data. They know that if they don't, their rivals most assuredly will.

Politicians understand science's peculiar power to convince, and so seek to appropriate its authority by claiming to create many of their policies in response to evidence. While they crave the power of evidence, however, too many of them fail to understand the responsibilities that go with it. Like Lansley and Smith, they make a show of using science in their judgements, but without properly grasping and drawing on what it is that makes it work. They are practised exponents of what Richard Feynman famously christened 'cargo-cult science'.

During and after the Second World War, several pre-industrial tribes in Papua New Guinea observed American soldiers clearing rainforest to build airstrips at their bases. This activity was inevitably followed by the arrival of planes carrying valuable cargo, such as medicine, food and clothing. After the Americans had left, some tribes sought to summon planes and their goods by building imitation airstrips of their own. These 'cargo cults' understood the value of the American technology, but misunderstood the true nature of how it worked.

Richard Feynman turned this story into a beautiful metaphor for pseudoscience. Cargo-cult science, he argued, similarly mimics the practice and terminology of science, without properly understanding the methods that give it such power to separate truth from falsehood.

Much alternative medicine is cargo-cult science. Homeopaths like to talk in technical jargon, they gravely measure out their doses with great precision and they quote studies that purport to show that their treatments work. Yet they do not subject themselves to the rules of science: they selectively quote from trials with positive results while forgetting the inconvenient negative ones, and they generally avoid the controls that are necessary for a worthwhile experiment. Their charade of science may pull the wool over the eyes of devotees, but it isn't evidence-based medicine.

Politicians who use spray-on evidence are also cargo-cult initiates. Perhaps because so few of them have a scientific training, or the lay geek's appreciation of how science works, they talk the language of evidence without understanding it, let alone heeding systematically what it has to say. When they claim to be guided by evidence, they are often in reality abusing and devaluing it. Science, as a House of Commons select committee put it, becomes 'at best a peripheral policy concern, and at worst a political bargaining chip'.

The many faces of evidence abuse

Evidence abuse comes in endless forms, none beautiful. There are so many species out there in the political wild, indeed, that it is possible to draw up something of a taxonomy.

Evidence-shopping. Jacqui Smith's approach to drugs classification was class A evidence abuse, which we can call evidence-shopping. Having already decided for political reasons that cannabis was to be upgraded, she shopped around for some science to go with the nice policy she'd picked up. She bought the recommendations of scientists who suited her, such as Robin Murray, of the Institute of Psychiatry, while rejecting those that didn't fit.

Imaginary evidence. Andrew Lansley took a different approach with his NHS reforms. With no real evidence available, he cited fiction. Imaginary evidence – or at least evidence that crumbles under close scrutiny – is pretty standard political fare. The usual assumption is that by the time it gets noticed, the policy will be in place, the news cycle will have moved on and the media's attention will be somewhere else.

Fixing the evidence. In 2001, President Bush announced the controversial embryonic stem cell policy (see Chapter 2), permitting federal funding of research on a few cell 'lines' already in existence, but blocking it for new ones. He justified it with imaginary evidence – his contention that the lines were sufficient was false, as most were not of adequate quality for medical research – and then compounded the error with another kind of evidence abuse.

Bush established a President's Council on Bioethics to advise him on such issues, and loaded it with members he could trust to deliver the kind of advice he wanted to hear. Its chairman was Leon Kass, well known as a conservative ethicist with strong objections to embryonic stem cell research. When one of the council's more eminent and independent-minded members, the future Nobel laureate Elizabeth Blackburn, began to dissent vocally from its opinions, she was sacked. This is fixing the evidence, manipulating the advisers so you get the right advice. We'll meet it again in Chapter 7.

Clairvoyant evidence. In May 2006, Patricia Hewitt, a Labour predecessor of Lansley's as Health Secretary, briefed the *Independent on Sunday* on her plans for a 'revolution in childbirth policy that will reverse decades of medical convention'. Convinced that home births were significantly less risky to mothers' health than much of the medical profession had previously assumed, and keen to widen choice, she would be ordering doctors to offer this option to every pregnant woman. The Department of Health had commissioned research to 'challenge the assumption that births should take place in hospitals', the newspaper reported.

Hewitt's confidence about the outcome of this research proved justified on this occasion: the following year, the Royal College of Obstetricians and Gynaecologists issued evidence-based guidelines that endorsed home birth as a good option for women with un-complicated pregnancies. But she put the cart before the horse. You can't know whether research will challenge assumptions until it's actually been done. Hewitt took a clairvoyant's approach to evidence, prejudging what it might say before it had been gathered. Jacqui Smith, a serial evidence abuser, did something similar in 2009 by announcing before the ACMD had completed a review that she would reject any advice to downgrade ecstasy's status as a class A drug.

Mishandling the evidence. Even when ministers use high-quality evidence that hasn't been cherry-picked, there's always a danger that they'll misinterpret it. Bovine tuberculosis is one of the most serious agricultural diseases in Britain: it led to the slaughter of 25,000 cattle in 2010 and cost the taxpayer £90 million. This form of TB is carried and spread by badgers, and in 2011, Jim Paice, the Agriculture Minister, ordered a badger cull to contain it. In support of his decision, the minister quoted evidence from a major, high-quality study commis-sioned by the government: the Randomized Badger Culling Trial.

A great example of evidence-based policy? Not quite. The trial did find that cases of bovine TB were reduced in 100 square kilometre zones where badgers were culled, but also that infection rates rose in the surrounding regions, as badgers carrying TB fled the cull. Mathematical models suggested that larger culling zones, of at least 150 square kilometres, would be needed to be sure of a moderate benefit, which might not be worth the cost of the cull. The evidence was not nearly as clear-cut as the minister suggested.

The misinterpretations didn't end there. In the randomized trial, badgers were trapped in cages before being culled – an extremely expensive approach. To reduce the cost, Paice instead allowed farmers to shoot badgers without trapping them first. There is no evidence that this would have the same effect. Indeed, there are reasonable concerns that this approach to culling could actually worsen the displacement problem as badgers flee the guns. The minister used evidence backing one policy to justify a completely different one.

Secret evidence. This is another political favourite. Soon after he became Education Secretary in 2010, Michael Gove abolished the £5,000 'Golden Hellos' that had been introduced to encourage science graduates to train as teachers, citing a 'value for money analysis'. But as the Campaign for Science and Engineering noted, this analysis hadn't been published. 'Where is it?' asked Imran Khan, the group's director. 'We're all for evidence-based spending, but there's no evidence that the Government has figured out what the impact of this cut will be.' Gove thus avoided the Lansley trap – if you keep your evidence secret, nobody can tell whether it's imaginary – while walking into another.

The limits of evidence

What all these forms of evidence abuse have in common is that they are employed to give a veneer of scientific justification to policies that are actually embarked upon for entirely different reasons.

Gove abolished 'Golden Hellos' when his department was being pressed to reduce spending. Paice backed the badger cull because farmers wanted one, and they represented an important Tory constituency that he wanted to appease. Bush had personal religious convictions that made him opposed to embryonic stem cell research. Smith wanted to escape a growing perception that Labour had been soft on crime in general, and on drugs in particular. Lansley was ideologically convinced that GP commissioning was the best way to transform the NHS for the better.

None of these motivations is improper. Ministers are entitled to take all sorts of factors into account when they reach decisions – indeed they must do so – and science will usually be just one of them. They are democratically elected; their advisers are not. When they weigh how to act, they are right to think about the expectations and aspirations of the people who voted for them, the promises they have made and the ideologies they espouse.

Tracey Brown, who campaigns for greater use of evidence in public policy as director of Sense About Science, is clear on its limitations. 'There's strong evidence that a 9pm curfew would cut crime, but it's clearly an infringement of civil liberties,' she says. 'It's as important to me that politicians are democratic as it is that they respect science.' Evan Harris, the former Liberal Democrat MP who has set up the Centre for Evidence-Based Policy since losing his seat, agrees: 'It's reasonable to expect questions of ideology, social justice and, yes, politics to feature in political decision-making.'

In his book *The March of Unreason*, Lord Dick Taverne, who founded Sense About Science, argued that politicians should leave independent experts alone to decide on fundamentally scientific issues, such as safe levels of pesticide residue in vegetables, public health measures such as vaccination schedules, or the best method of storing radioactive waste from nuclear power plants. He's right that ministers and civil servants without specialist expertise have little of note to contribute to the technical aspects of such questions. But while the best scientific evidence is of paramount significance to this type of decision, and ministers need exceptionally good reasons for rejecting it, it remains important even here for democratically elected representatives to have the final say.

It is exceptionally rare for scientific evidence to mandate a single solution to a policy problem; rather, it informs the range of solutions that might be feasible and predicts what the outcome of each is most likely to be. Questions of ethics, law, public acceptability, fairness, personal liberty and economics are usually relevant too, and scientific experts are often no better placed than anyone else to judge these. If we value democracy, advisers should advise and ministers should decide.

The degree to which pesticide residues potentially damage health is a scientific question which science can answer. Science can also help to explain whether the higher cost of vegetables that would result if pesticides were more strictly controlled or banned would lead to better or worse overall public health. It can't, however, decide how society should balance risks and benefits: that is a political question for elected representatives to take.

Science can tell you how requiring children to present an immunization certificate to attend primary school, as happens in many US states, would improve herd immunity, but not whether this is an acceptable infringement of parental rights. On drug control, global warming, embryonic stem cell research and abortion, science can describe the challenges and the likely effects of the possible solutions. The value judgements of democratic societies, however, matter too.

Few civilian politicians have much knowledge of how to drive a tank, to fly a Typhoon strike jet, or to plan a military campaign. Yet we do not leave decisions about warfare to army officers: in democracies, we expect them to take their lead from elected governments. Science shouldn't expect to be treated differently. Geeks are as important as generals in the specialist advice they have to give, but they aren't more important.

Yet this military analogy also cuts in another direction. While we accept that the buck stops with the president or prime minister, not with the chief of defence staff, we do not expect civilian leaders to order the armed forces into battle without first consulting them fully and frankly about the feasibility of action and its possible outcomes. Neither would we expect them to start a war that generals think unwinnable, or to overrule their advice on how best to pursue it without very good reasons. As Tim Harford describes in his book *Adapt*, the US military had to counter Donald Rumsfeld's political neuralgia about the very existence of an 'insurgency' to establish the counter-insurgency strategy that ultimately stabilized Iraq. Had the Secretary of Defense been prepared to take a little more specialist advice from officers on the ground, it is arguable that many American and Iraqi lives would have been saved.

So it is with science. Ministers aren't obliged to make every decision according to the evidence presented to them by scientists and nothing else. They should, however, ensure that they do take scientific advice on questions to which it is pertinent – and there are few matters of public policy to which it is not. Evidence isn't usually sufficient for sound policy-making. But it is nearly always necessary.

Advice from scientists with relevant expertise should be sought and considered in good faith before decisions are made, rather than sprayed-on afterwards. It should be published in full (excepting rare cases in which national security or commercial confidentiality precludes this) and advisers should not be admonished, silenced or sacked for explaining their conclusions in public. If ministers decide to overrule expert advice, as they are entitled to do and often will, they should explain their reasons. When they choose to go with instincts that point one way, over evidence that points another, they should say so.

Above all, politicians and civil servants should not be allowed to get away with laying claim to evidence-based policy when decisions have actually been taken by other means. 'I don't think there are very many scientists who are naïve enough to think that science should always determine outcomes,' said John Holdren, before becoming President Obama's chief scientific adviser. 'But you shouldn't defend outcomes by distorting the science.'

A better appreciation of science and its methods among politicians would breed a healthier respect for the currency of evidence, which is devalued a little with every act of abuse. It might also start to persuade more politicians that they need robust, high-quality evidence at an early stage, and that they could draw more on the methods of science to devise better policies that really deliver. There isn't just a need to change the use of evidence by government. The way it is collected needs to improve as well.

Wilful blindness

Iceland, the world's eighteenth largest island, sits in the North Atlantic at the junction between the Eurasian and North American plates. It has existed only for about 20 million years, formed by a series of volcanic eruptions that heaped lava flow upon lava flow to raise it from the ocean floor. It is among the most geologically unstable places on earth, with twenty-two active volcanoes. When one of the smaller ones, Eyjafjallajökull, began to erupt spectacularly on 20 March 2010, it shouldn't have been surprising. But it caught European governments on the hop.

As the volcano spewed a huge plume of volcanic ash into the sky, the wind carried it south towards Europe. Volcanic ash presents a serious hazard to aviation – it can cause aircraft engines to fail – and regulators' guidelines recommended closing airspace as soon as concentrations rose above zero. For two weeks from 16 April, much of Europe became a no-fly zone as governments acted accordingly. The European Union estimated the cost to airlines alone at £2.2 billion.

These draconian and ruinously expensive safety measures were founded on very limited evidence: while the threat from thick clouds of ash was well understood, the dangers presented by the much lower concentrations present over Europe were much more uncertain. As test flights assessed the situation and engine manufacturers started calculating, it soon became clear that aircraft could fly safely through thin clouds of ash. The safety thresholds were raised and airspace re-opened, but not before immense economic damage had been done.

Politicians and regulators had little choice but to respond to the ash crisis as they did: just imagine for a second the outrage had a Boeing 747 gone down. That the eruption caused a crisis at all, however, was largely of their own making.

Major volcanic eruptions in Iceland are not exactly rare, and while the weather system that carried the ash cloud over Europe was atypical, the island lies close to transatlantic flightpaths. The need to plan for an Eyjafjallajökull-type event, and to work out safe thresholds for flying, could and should have been anticipated. Volcanologists and air-traffic controllers had been urging governments and the airline industry to do so for years. They sat on their hands.

A year after the crisis, a damning report from the Commons Science and Technology Committee found that the threat from volcanic ash had not been considered in Britain's National Risk Assessment exercise, despite warnings from scientists. Sue Loughlin, head of volcanology at the British Geological Survey, told the committee: 'It wasn't particularly a surprise to the volcanology community that something like this would happen, but somehow that message hadn't got through to Government.' It turned out that the government's chief scientist, Sir John Beddington, hadn't even been consulted on what the National Risk Assessment should include. A failure to ask scientists to evaluate an entirely foreseeable threat contributed directly to losses worth billions of pounds.

'We have been left with the impression that while science is used effectively to aid the response to emergencies, the Government's attitude to scientific advice is that it is something to reach for once an emergency happens, not a key factor for consideration from the start of the planning process,' the select committee concluded. 'This is not trivial: had the risks of volcanic ash disruption been assessed prior to the emergency, the Government would undoubtedly have been better able to cope with the situation.'

Politicians can't be expected to act on evidence that doesn't exist, and they are sometimes required to take quick decisions that can't wait for new research. But as the ash crisis shows, they are often part of the reason why the evidence is absent. Systematic recourse to scientific advice in every aspect of government can help to establish gaps in our knowledge that might be remedied by commissioning research. That's research, by the way, of the genuinely open-ended variety, not the type demanded by Patricia Hewitt to prove a point.

This reluctance to gather robust evidence is particularly acute where politicians' pet policies are concerned. For while their architects don't like to admit it, all new initiatives that governments put into action are to some extent leaps in the dark. Whether they are founded on strong and genuine evidence, or have no evidence base at all, not even the brightest experts know quite how things are going to turn out. Policies are experiments. Many of them – perhaps even a majority of them – will not achieve what their authors intended. 'Half of all business ideas fail, and 10 per cent of companies fail every year,' Tim Harford points out. 'Why should we assume that politics does any better?'

This shouldn't matter. In fact, it should really be something to celebrate. Experiments are the stuff of science, a powerful route by

'Half of all business ideas fail, and 10 per cent of companies fail every year. Why should we assume that politics does any better?' – Tim Harford

which we can find out whether interventions really work. Evolution can be understood as a series of natural experiments, in which genetic variations are put to the test by the environments to which they're exposed. In *Adapt*, Harford makes a compelling argument that the trial and error of experiments is the surest route to solving all manner of problems, from counter-insurgency in Iraq to development in Africa. Trying out lots of ideas and developing the good ones while discarding the bad is how science builds knowledge and understanding about the world. 'All life is an experiment,' wrote Ralph Waldo Emerson. 'The more experiments you make the better.'

For experiments to be valuable, however, their experimental nature needs to be recognized. When it comes to their policies, this tends to pass governments by. Unable to accept that new drugs laws, school systems and pollution regulations are experiments, they fail to learn from them as they could. As we'll see when we look in more detail at education and crime, if trials or pilots are conducted at all, they're often designed badly. Once policies roll out nationally, they rarely get the systematic and dispassionate evaluation that experiments need if we're to find out whether they really worked.

This happens in part because too few politicians and civil servants are well enough versed in science to understand the value of experiments. It also reflects two troublesome realities of today's politics.

The political cycle means that the minister who launches a policy will rarely be the one who sees things through. If he commissions a pilot, the chances are that by the time it's reported back in a year or two, he'll have been promoted if he's done well, or sacked if he hasn't.

What is more, there's actually a disincentive to subjecting your policies to independent and dispassionate scrutiny. Evaluate that knife-crime initiative really rigorously and there's a good chance that the data will show that it failed. The opposition will have a field day. You'll be the minister who wasted millions on a pet project that didn't work. Political toast. Far better to commission some half-baked qualitative analysis that anyone can interpret as they choose. It might not settle the argument in your favour, but neither will it do you in.

Up to us

We can't just blame politicians for this intolerance of error. As voters, we are all at fault. If we weren't so quick to scorn the architects of mistaken initiatives undertaken in the spirit of experiment, they might become keener to appraise their progress properly as they innovate.

As geeks become more politically active and involved, this is an area where we can help. When a scientist proposes a hypothesis that fails to be confirmed by the data, there may be disappointment but there is no dishonour, so long as he accepts the outcome and moves on. We should praise honest failure accomplished by applying the scientific method to public policy and encourage U-turns made in the face of the evidence.

When she first introduced draft legislation in 2006 that would become the Human Fertilisation and Embryology Bill, Caroline Flint, the Health Minister, proposed banning the creation of embryos containing human and animal material. When virtually every scientific organization of influence criticized the damage this would inflict on research, she changed her mind. This mature approach to policy-making deserves to be celebrated. We must resist the temptation to score points against politicians who backtrack after weighing the evidence; we should congratulate them instead.

Harford has suggested awarding an annual prize to the politician who has been most usefully wrong and admitted to it. That would be a great start. We could even go further, actively demanding that candidates give us examples of mistakes from which they have learned. To establish whether a politician appreciates science at hustings, the question 'What have you been wrong about?' is more useful than testing their knowledge of Newton's laws of motion. As we build online resources to identify candidates who deserve our votes, we should note with approval those with a record of testing political hypotheses and abandoning the failed ones. Equally, those who claim nothing but brilliant ideas deserve a red flag.

We must also name and shame the evidence abusers. Politicians promote policy-based evidence not only because few of them properly understand how to handle data, but also because they know they can get away with it. Once again, we have to play our part in changing this.

Many geeks are adept at finding, checking and interrogating evidence, using the many online databases that collate the peer-reviewed literature and make it searchable. Medical research is indexed by an online archive called PubMed, and the Cochrane and Campbell Collaborations prepare and disseminate systematic reviews of medical and social science respectively, establishing what all the good-quality research in a field says when assessed as a whole. The Freedom of Information Act allows us access to much of the evidence on which policy decisions are founded. We have the skills to test political claims forensically against these data.

Through blogging and social networking, we also have the tools to share what we discover. Web-based initiatives such as Straight Statistics, FullFact.org and Channel 4 News's FactCheck have opened new forums for exposing dodgy claims. Straight Statistics, for example, has demolished a string of misleading government announcements about crime rates. When evidence abuse is revealed, the mainstream media will often take up the story: it was scientists such as Lord Krebs who blew the whistle on the bad science behind the badger cull. It has never been easier to fact-check politicians and embarrass those who commit the sins we met above.

Evidence abuse is a form of what the journalist Peter Oborne calls political lying, and politicians never look good when their lies are found out. They need to understand that they're being watched, that if they claim that a policy is evidence-based, someone is going to check the references. Only then will they begin to learn to take the self-critical approach that is ingrained in science, ensuring their data are suitably robust before they go public with them, for fear that if they don't verify their conclusions properly somebody else most surely will.

We can exploit the political system to our advantage here. Most MPs and congressmen might not be schooled in science, and are not averse to a bit of evidence abuse, but they also understand that spotting it in their opponents and rivals can be a fruitful way of advancing both their agendas and their careers. They also have powers that ordinary citizens lack. They can ask parliamentary questions that ministers must answer. Many sit on select or scrutiny committees that hold the executive to account. They have the profile and platform to embarrass an evidence-abusing secretary of state. If they're ambitious, they'll be looking for issues on which to make a name for themselves and build political capital, so they catch the eye of the party whips. They also represent us, and at least some of them will be willing to help us if we help them.

'You should see your MP as somebody who gives you the ability to hold bigger and more powerful people than yourself to account,' says Tracey Brown. 'Bringing issues to their attention isn't necessarily lobbying. You can help them out, feed them information they can use.'

If our representatives lack the skills or inclination to spot evidence abuse for themselves, we need to put it in front of their noses and challenge them to do something about it. Invite them to use our expertise. Become a useful resource. Some of them will take up our causes, in forums that ministers really notice. And we'll also convey the implicit message that evidence matters to more people than they might have realized, and that if they slip up then the geeks will be on their case.

There are other opportunities to participate actively in the political process that we need to seize when they present themselves. When governments start to draw up new legislation, they usually begin with a consultation document that sets out what ministers are minded to do and invites comment from any interested parties. Some consultations are more genuine than others, but it is critical that those of us who value scientific thinking respond in numbers – and that private individuals contribute as well as organizations. Other groups with whom geeks often disagree, such as the embryo rights lobby and Greenpeace, are highly effective at mobilizing their supporters at consultation time. Science doesn't do this nearly so well.

When the Department of Health ran a consultation before drafting the Human Fertilisation and Embryology Bill, it received 535 responses. While scientific and medical organizations took part, few individuals from this community shared their views. Anti-abortion groups also prepared submissions, but rallied dozens of individual supporters to contribute as well. The result was that views collected from 'ordinary people' were skewed against liberalizing the law, particularly on the controversial issue of human–animal hybrid embryos. Ministers formed the mistaken impression that the public was queasy about this research, and were spooked into framing restrictive legislation that took a long campaign to unwind.

Parliamentary and congressional committee inquiries offer another chance to take issue with evidence abuse in a way that can make a difference. Yet as with consultations, too many geeks who would have something useful to say remain silent: they either miss the 'call for evidence', or they lack the time or motivation to respond.

Dave Wark, the physicist we met in the last chapter, started to contribute to select committees only after taking part in the Royal Society pairing scheme, testifying on physics funding and student visas. 'I was surprised by how little input they get,' he says. 'I'd always imagined that the pros would deal with it, but I guess I am a pro. It left me asking why I'd not done it before. If people like me don't contribute what we know, nobody's going to do it for us.'

It would help enormously if science had more of an organizational base, with mass-membership groups that can mobilize for inquiries and consultations and coordinate action. The Campaign for Science and Engineering and Sense About Science are effective lobby groups, with websites that offer helpful advice about how best to influence policy-makers. But both run on tight resources. CaSE has only 925 individual members. If you're interested in evidence-based policy, you should really sign up.

Another group, Science is Vital, has started to exploit the database of 45,000 supporters it built up when campaigning to protect the science budget. In the summer of 2011, it emailed its subscribers inviting them to share their concerns about the structure of science careers with David Willetts, the Science Minister. Evan Harris hopes his Centre for Evidence-Based Policy will perform a similar co-ordinating role, highlighting the consultations and inquiries to which we should be contributing.

As this book went to press, the National Endowment for Science, Technology and the Arts, a lottery-funded body that promotes innovation, was establishing a UK Alliance for Useful Evidence. It aims to encourage the collection of rigorous data that can inform policy, particularly in fields such as education and criminal justice where it is often lacking.

The basis for more concerted action is there if we give these groups the support they need.

'We need to use the opportunities that events provide us with to make the use of evidence into a central theme of political debate,' says Harris. 'It's hard to predict what the issues we can fight on will be, but we need to grasp them when they arise. We need to establish the political narrative that science and evidence matter in politics, and that plenty of people care about them.'

There's only so much we can do from beyond government, however, to promote better use of science and evidence. We also need to make the case that the structure of government itself must embed these values into policy-making.

Geeking government

Every UK government department now has a chief scientific adviser: the Treasury was the last to appoint one, in June 2011. But they vary widely in their power and influence.

Some, such as Dame Sally Davies at Health, have substantial budgets and sit on the departmental board. This gives them a critical voice in decisions over the organization's direction of travel. These, though, are departments in which evidence tends to be respected (if not as much as it could be), because they oversee aspects of government in which technology and evidence can make the difference between life or death.

In other departments, such as Transport and the Home Office, the chief scientist is a more marginal figure, whose input is not seen as central to most of the decisions it makes. The Treasury chief scientist is an economist who continues to do his old job in tandem, hardly a sign that the most powerful department of all takes his contribution especially seriously. The Campaign for Science and Engineering has produced a 'scorecard' that marks the influence of chief scientists according to six factors, such as the frequency with which they meet ministers and their control over research budgets. No department scored six and only one, the Department for Environment, Food and Rural Affairs (Defra), scored five.

These institutional structures matter. What Bob Watson, chief scientist at Defra, values most is his seat on the departmental board. 'It means I'm part of all the key discussions,' he says. Giving every chief scientist similar status would immediately raise the profile of science in policy-making, without financial cost.

'Chief scientific advisers need to have policy sign-off powers,' says Chris Tyler, director of the Centre for Science and Policy at the University of Cambridge. 'Just as chief economists sign off every policy as economically viable, and lawyers look at the legal implications, someone needs to take a look at how the evidence has been used.'

This wouldn't mean replacing political choices with technocratic and undemocratic ones: ministers would remain in charge, and at liberty to overrule their advisers. But such a system would at least ensure that science and evidence are properly considered, even if the ultimate decision is that other factors outweigh them.

Geeks also need to develop a better understanding of how policy is really made, to learn their way around the machinery of government. 'The geek movement is often more comfortable outside the tent pissing in,' says Tyler. 'Sometimes, things can only be changed from the inside.' That means engaging and building relationships with the civil servants who frame the detail of most government policies and put them into action.

The civil service and the scientific world rarely have much experience of one another, which benefits neither. Officials are unaware of research that might be relevant to their fields, and scientists uncertain how best to offer their insights to those in charge of developing policy. Professor Ben Martin, of the Science and Technology Policy Research Unit at the University of Sussex, argues that there's a need for more movement between the academic, industrial and civil service sectors, and exchange and sabbatical schemes have much to contribute here. Civil servants could benefit greatly from spending time in universities to make contacts and learn from the way evidence is used. The Centre for Science and Policy at Cambridge has begun a 'Policy Fellowship' scheme to achieve this, which could be emulated elsewhere.

Virtually every university now has a knowledge-transfer office for industry, staffed with people whose job it is to build links between scientists with bright ideas and businesses which might exploit them financially to the benefit of all parties. Virtually none has a similar body that helps scientists and social scientists whose work has implications for public policy.

By the same token, science funding arrangements could do more to reward researchers who contribute their expertise. At present, there are too many disincentives: providing expert advice is time-consuming and takes scientists away from the research on which their grants and career prospects ultimately depend. More scientists could be seconded to government to learn how policy-making functions. Young scientists, in particular, should be offered opportunities to engage with the ways of Whitehall. 'About 60 per cent of PhD students don't go on to do postdocs,' says Tyler. 'Let's ensure that if they're funded by the taxpayer, through the research councils, they get some training in policy. It would be dirt cheap to do, and it would get them involved. If they go on to do science, they'll understand the system. And some of those who don't might consider the civil service as a career.' The civil service could do with a few more geeks.

Scientific responsibility

When George Osborne became Chancellor in 2010, one of his first decisions was to establish an Office for Budget Responsibility (OBR). He gave this independent body a statutory duty to scrutinize the state of the public finances and the Treasury's economic forecasts. Osborne's goal was to give both himself and his successors an incentive not to fiddle the figures for political reasons. Executive power over the economy remains, as it should, with the Chancellor and his Treasury ministers, who are elected; but if a Chancellor makes calls on the basis of over-optimistic or otherwise questionable economic data, there will be a price to pay.

If this can work for economics, why not for science? It is no less likely than economics to be abused for political reasons, and regular audit would be a fine way of keeping ministers honest. An Office for Scientific Responsibility could scrutinize the data advanced in support of every new government policy and report on whether it really passes muster. It could also examine the steps taken to ensure that evidence of the effects of new initiatives are properly collected and evaluated. Ministers would still be free for political reasons to embark on actions that lack a sound evidence base, but if they claimed evidence that does not exist, or failed to test the policies they introduce, they could expect an official rebuke.

A system rather like this was in fact proposed by the Commons Science Committee in its 2009 report, *Putting Science and Engineering at the Heart of Government Policy*. Departmental chief scientists, it recommended, should press ministers and officials to show their working, or to acknowledge that their decisions were not evidence-based, and the government chief scientist should then name and shame those who refuse in an annual report. 'We are not saying all policymaking must be based on clear evidence,' said Phil Willis, the committee's chairman. 'We fully accept that governments come in on a party manifesto and take into account a whole host of other criteria when making policy. We have no problem with that. We do have a problem with claiming there's evidence they don't have.'

The departmental chief scientists are probably not the right people to do this job. Both Sir John Beddington and Bob Watson consider that having to conduct such a naming and shaming exercise would risk undermining their relationships with ministers and senior civil servants, diluting their day-to-day influence accordingly. But an independent Office for Scientific Responsibility would not have this drawback. It would take political bravery for a government to hamper its own room for manoeuvre in this way, and to create a censure system of which it is bound to fall foul before long. As Osborne's creation of the OBR shows, though, such bravery does exist, and his decision will be difficult for any successor to unwind. A similar body for science could institutionalize respect for evidence across government and force ministers who lack it to account for their behaviour, and thus to pay a political price.

We can't bank on new legislation like this, and we certainly can't wait for it. If policy-based evidence is to be rooted out of politics, it's incumbent on us geeks to get the ball rolling now. We can use carrots, offering our expertise freely and openly to those ministers, MPs and civil servants who are willing to accept it. But we also need sticks, effective ways of convincing politicians that abusing evidence carries a cost.

One of those sticks has to be the media. It's the sea in which politicians swim, exerting great influence over their positions and priorities. Yet as we're about to see, the way it approaches science could use some geek action too.

UNFAIR AND UNBALANCED

Why science matters in the media

ON 28 SEPTEMBER 2009, A FOURTEEN-YEAR-OLD GIRL COLLAPSED AND died at her school in Coventry. Natalie Morton, an autopsy showed, was killed by a large chest tumour that had not been detected. It was a tragedy that made the headlines, but not because of the missed diagnosis. On the day of her death, Natalie had been given a human papilloma virus (HPV) vaccine called Cervarix under a new national programme to immunize girls against an infection that is the primary trigger for cervical cancer.

The coincidence – and it was nothing more than that – was sufficient for much of the media to indulge a favourite preoccupation and whip up a health scare. 'The sudden death of Coventry schoolgirl Natalie Morton after a jab against cervical cancer highlights the reality that vaccination programmes are not without their risks,' wrote Richard Halvorsen, a GP critical of many vaccines, in the *Daily Mail.* 'The tragic irony for Natalie was that the injection may have triggered a reaction far more lethal than any future, distant threat of a comparatively rare disease.'

As Halvorsen's article was published, an inquest was hearing from Caron Grainger, director of public health for NHS Coventry. 'There is no indication that the HPV vaccine . . . was a contributing factor to the death, which could have arisen at any point,' she said. Natalie was not a 'cancer jab victim', and her case had nothing to say about vaccine safety. Doctors, the coroner and Natalie's parents all accepted that there was no link. Yet in spite of that, sections of the media that really ought to know better elected to prolong this non-story.

The morning after the inquest, the *Today* programme on BBC Radio 4 held a discussion between Adam Finn, Professor of Paediatrics at the University of Bristol, and Halvorsen. Though it was now known that Natalie's death was not caused by Cervarix, Halvorsen was given a platform to claim that 'one has to be still suspicious that the vaccine could at least have been a trigger for the death', and to assert more widely that the benefits of immunization had been exaggerated and the risks downplayed.

As Finn pointed out, Halvorsen's views lie well outside the scientific mainstream and his analysis of Natalie's death in the previous day's *Mail* had been proved completely wrong. That did not stop the BBC from inviting him on to its most influential current affairs programme to raise groundless safety concerns that could influence people's behaviour, in the name of lively debate and balanced coverage that allows both sides their say.

It was a discussion that encapsulates the worst aspects of the media's approach to science, which serves to confuse and mislead the public and to damage the status of science in society. It demonstrates how an insatiable editorial appetite for arresting headlines and talking points leads to the distortion of research and to the invention of fictional scientific controversies. The equality of prestige accorded to Finn, who was representative of the views of the overwhelming majority of doctors, and Halvorsen, who was not, is also typical of a phoney balance that further misrepresents science and medicine.

The balance fetish

The media thrives on controversy and novelty, and rightly so. The business of news is drawing an audience through reporting the new, the important and the interesting, and that will often involve focusing on contentious subjects that elicit strong views and polarized debates. Media culture, too, particularly in its broadcast form, also places great value on balance. Many journalists consider it their duty to tell both sides of a story, and a commitment to 'due impartiality' is enshrined in the BBC Charter and its editorial guidelines.

This works pretty well for media coverage of many walks of life. If a government minister is interviewed, the opposition spokesman can be given equal time to make her case. A complaint from a consumer group about a faulty product can be put to its manufacturer for comment. This simplistic formula, however, is often inappropriate for science. By making a fetish of balance, and insisting too rigidly that both sides of a story are told, it becomes very easy to mislead.

In science, opposing views do not always have equal merit. One side of an argument is often well supported by peer-reviewed evidence, while the other has little or no such backing. It might be balanced to accord similar status to each position and always to allow both the right of reply. But it isn't necessarily fair to the science.

As the science writer Chris Mooney put it: 'The journalistic norm of "balance" has no parallel in the scientific world, and, when artificially grafted on to that world, can lead reporters to distort or misrepresent what's known, to create controversies where none actually exist, or to fall prey to the ploys of interest groups who demand equal treatment for their "scientific" claims.'

Steve Jones, the geneticist, reached a similar conclusion in a review of the BBC's science coverage he led in 2011. 'Bogus impartiality (mathematician discovers that $2 + 2 = 4$; spokesperson for Duodecimal Liberation Front insists that $2 + 2 = 5$; presenter sums up that "$2 + 2 = $ something like 4.5 but the debate goes on") can, perversely, lead to bias in its own right, for it gives disproportionate weight to minority views – and some of the minorities involved are expert in taking advantage of the platform offered.'

One of the worst examples of this distortion concerned another vaccine: the triple jab against measles, mumps and rubella (MMR). When Andrew Wakefield, of the Royal Free Hospital in London, suggested in 1998 that the MMR vaccine might cause autism and advised parents to use single vaccines against the three diseases instead, he had very little evidence to support his claims. The link was not even proposed in his infamous case series report in the *Lancet*; Wakefield raised it in a press conference, where his comments were unconstrained by the demands of peer-review. Barely any other doctors and scientists shared his view that MMR posed dangers, beyond the very rare side effects that all vaccines can have, and their confidence was backed by epidemiological research.

Much of the media, however, including parts of the BBC, presented Wakefield's unfounded and now discredited assertions as if they were one side of a genuine medical debate and not the outlandish views of a single maverick. Opponents and defenders of the vaccine were routinely pitted head to head. As Tammy Boyce, of Cardiff University, describes in her book on the saga, *Health, Risk and News: the MMR Vaccine and the Media*, the impression given was of genuine dissension between experts. Worried parents were left confused about whom to believe.

'Mathematician discovers that 2 + 2 = 4; spokesperson for Duodecimal Liberation Front insists that 2 + 2 = 5; presenter sums up that "2 + 2 = something like 4.5 but the debate goes on."'
— *Professor Steve Jones*

While some specialist health and science journalists, such as Nigel Hawkes in *The Times* and Sarah Boseley in the *Guardian*, robustly criticized Wakefield's claims, other reporters were drawn to the scare by fatal attraction. They played it up for everything it was worth, defying the evidence in pursuit of a good story. Heart-rending anecdotes from parents who blamed MMR for a child's autism were set against reassurances from paediatricians made out to appear complacent or heartless. The prevailing attitude was betrayed at the British Press Awards in 2002: one of Wakefield's most loyal newspaper cheerleaders, Lorraine Fraser of the *Sunday Telegraph*, was named Health Writer of the Year by her industry peers.

Fair and impartial reporting of the MMR scare, Boyce explains, 'would not have been balanced, because, in reality, the evidence was not balanced'. The decision by many journalists and editors to ignore this magnified a shoddy piece of science to the point where it inflicted real damage to public health. In 1998–9, when the vaccine scare first surfaced, 88 per cent of two-year-olds in England had been immunized against measles, mumps and rubella; by 2003–4, coverage had fallen to 80 per cent, and as low as 61 per cent in some parts of London. Measles made a corresponding comeback: there were fifty-six confirmed cases in England and Wales in 1998, and 1,370 in 2008. In 2006, a thirteen-year-old boy from a Traveller community became the first child in Britain to die of measles in fourteen years.

If vaccine scares supply the most egregious examples of phoney balance, they are far from the only ones. It has affected debate over climate change, as the contrarian views of the small minority of scientists who do not accept that human activity is driving global warming have attained a media profile out of all proportion to their scientific credibility. Genetically modified crops have been labelled as dangerous 'Frankenfoods', despite overwhelming evidence accepted by a large majority of scientists that they are safe to eat. Reports of health risks from power lines and mobile phones, and of the supposed benefits of alternative medicines, are similarly distorted. Embryo rights and anti-vivisection campaigners are not only invited to question the ethics of stem cell research and animal experiments, on which they deserve a fair hearing; they are also permitted to challenge their utility, without presenting much evidence.

Reporting both sides is fair enough, good practice even, when both sides can call on decent data to support their positions, or when the evidence can reasonably be interpreted in multiple ways. That's why it works well in politics, where, as we've seen, policies are more usually founded on ideology, populism or personal hunch than on a thorough examination of all the available evidence. It is neither necessary nor desirable when the evidence is clearer.

Good science journalism needs a broader sense of impartiality, founded not on knee-jerk 'balance' but on critical appraisal of competing claims. The goal should be not to seek out contrary views for their own sake, but to include those that are robust and relevant. That means acknowledging the uncertainties inherent in science and reflecting reasonable dissent from the mainstream. But it also means showing proper scepticism towards arguments that are lacking in evidence, scientifically implausible or plain wrong.

Everybody has the right to an opinion about science, but some of those opinions are more equal than others.

Bad science

The media's abuse of science is not confined to bogus impartiality. Good news as well as bad is liable to be hyped, with new findings about disease exaggerated as 'breakthroughs' or 'cures' when few deserve that accolade. Science more usually advances in incremental and faltering baby steps, and encouraging preliminary results regularly disappear on further analysis. Almost all medical advances begin with observations of interesting biological effects in cultured cells or in animals, but the majority of these observations eventually lead nowhere. You could often be forgiven for missing that, though, from reading the headlines. Many 'cancer breakthrough' stories are indeed good news, so long as you're a mouse.

As this chapter was being written, in April 2011, the *Daily Telegraph* ran a front-page splash under the headline: 'New heart attack jab even more effective than statins'. It described a 'simple injection given to patients up to twelve hours after a heart attack or stroke', which 'could reduce their devastating effects by more than half'. It took the paper eleven paragraphs to tell its readers that this wonder drug had so far been tested only on animals, and that human trials would not begin for at least two years. The story wasn't inaccurate in its details, but its language and front-page position were profoundly misleading.

Breathless reporting like this can raise false hopes: desperate patients with multiple sclerosis have squandered money on worthless stem cell treatments that lack evidence of effectiveness, yet which have been promoted by journalists as miracle cures. It also feeds a widespread sense of cynicism about science, which emerges when fulsome reports of encouraging preliminary results are not borne out by subsequent trials. The perception forms that scientists are not to be trusted, that they will claim anything for a little publicity that might boost their status or funding, that the benefits of their work have been oversold. Sometimes scientists are indeed guilty of such hype, but it is more usually the media that does the overselling.

In his book *Bad Science*, Goldacre presents a catalogue of further ways in which the media misleads. There is credulous reporting of wacky but inconsequential science that isn't science at all, but apes it for the purposes of public relations. Think dodgy 'scientific' formulas for the gloomiest day of the year, the perfect penalty kick, the ideal handshake. Then there are the repeated stories that suggest scientists cannot make up their minds about whether coffee, red wine, chocolate and the like cause or cure cancer.

Flagrant misunderstanding of statistics is routine, as is presentation of risk stripped of its context. In January 2011, Channel 4 News ran a scoop on pregnancies among women using a contraceptive implant called Implanon. It failed to balance scary-sounding headlines about unwanted pregnancies with statistics that show the implant to be significantly safer than other methods of contraception – all of which have acknowledged failure rates. An effect that could have been entirely normal was portrayed as a serious problem. Such coverage can often have highly deleterious effects. In 1995, new research linked an old type of contraceptive pill to a raised risk of deep-vein thrombosis. Though the absolute risk of a fatal thrombosis was low – about two deaths per million women – the media scare led thousands of women to abandon their contraception, and to 10,000 extra abortions.

The overall effect is to mislead ordinary people not only about the findings of science, but about the discipline itself. 'In their choice of stories, and the way they cover them, the media create a parody of science,' Goldacre argues. 'On this template, science is portrayed as groundless, incomprehensible, didactic truth statements from scientists, who themselves are socially powerful, arbitrary, unelected authority figures. They are detached from reality; they do work that is either wacky or dangerous – but either way, everything in science is going to change soon – and, most ridiculously, "hard to understand". Having created this parody, the commentariat then attack it, as if they were genuinely critiquing what science is all about.'

The picture Goldacre presents is itself parody to an extent. His book highlights the many examples of poor science reporting to make a justified point, but there is plenty of good stuff out there as well. Most science and health specialists, at least at the more serious end of the newspaper market, make a genuine effort to convey medical and scientific developments accurately and in context, within the difficult confines of short items that must be accessible for a general audience. They succeed more often than they fail. It isn't fair to damn all journalists according to the failures of the bad ones, or to present the mainstream media as if it were a monolithic whole – any more than it would be to judge all medical researchers by the standards of Andrew Wakefield. The very word 'media' is a plural noun, and there is plenty of plurality within it. Different newspapers, TV and radio programmes and websites take very different attitudes to science, as do the journalists who work for them.

None the less, Goldacre's critique captures an important truth. A doctor by profession, he makes the sound diagnosis that many of the media's problems with science are cultural in nature, brought about by the dominance within it of humanities graduates and a shortage of senior figures with an instinctive appreciation of science and its methods. Just like politics, the media is not steeped in science and doesn't generally value it the way it should. In short, it suffers from a shortage of geeks.

A geek-free zone

All the major UK media outlets report regularly on science. Every serious paper and news programme has at least one science correspondent, and the BBC has a couple of dozen. Both the volume and the quality of the coverage can be said to be improving. A 2009 study led by Andy Williams, of Cardiff University, found that the number of science, health and environment reporters almost doubled between 1989 and 2005, rising from 43 to 82.5 (the half represents a part-time post). It also reported a growing appetite for science among news editors, which is reflected in a handful of exciting new initiatives.

The Times, as we've seen, launched *Eureka*, a glossy monthly science magazine, in 2009. The following year the *Guardian* began hosting a thriving science blog collective and its Sunday stablemate, the *Observer*, started a four-page science section called *Discover*. BBC One launched *Bang Goes the Theory*, a weekly science magazine series, in 2009; the channel had been without a popular science show since *Tomorrow's World* was cancelled in 2003. The BBC declared 2010 to be its 'Year of Science', increasing the volume of science across its output and screening the *Wonders of the Solar System* series that started to turn Professor Brian Cox into a household name. So much so, indeed, that the physicist was signed up to write a regular column for the *Sun*.

Yet despite these positive indications, science remains largely peripheral to the media's core business and values. Yes, it has a place at the table. But it sits below the salt, unlike other aspects of public life with which editors are more comfortable: politics, business, sport and the arts. Science is seen as a niche interest, not as a central part of the national conversation.

In news pages and bulletins, it features principally in the context of medical advances, health or environmental concerns, or controversies, which, as Goldacre notes, are often reported with little regard to the underlying evidence. Alternatively, it is light relief, or a fascinating diversion from meaty but often bleak stories about world affairs, conflict, economics and politics.

When it is the subject of popular documentaries, challenging content is often stripped out: members of the BBC's award-winning Natural History Unit complained to Steve Jones that they were warned to avoid too much about ecology or evolution, so they made 'children's programmes, albeit superbly good children's programmes'. Rarely is science portrayed as an integral part of society, as a theme that cuts across political, social and cultural issues in the manner of economics, law or the arts. It lives in a ghetto. The ghetto might be growing a little larger, but its walls remain.

You can see this when you compare the way newspapers handle science and business or the arts. Every paper has a daily business section, and dedicated arts sections that run at least weekly. *Eureka*, at *The Times*, is the only stand-alone science supplement in a British newspaper. The business and arts departments have a staff of specialists with different beats. The film critic is not expected to review exhibitions at Tate Modern, nor the literary editor to cover pop music. Science is not accorded similar status. Specialist science reporters must handle everything from particle physics to DNA. And they rarely hold senior positions with responsibility for ensuring that the whole staff, from the features desk to the sports team, gives scientific issues the accurate and responsible coverage they deserve.

None of the editors of the UK's national papers has a degree in science or a background in reporting it. Such people are also in short supply among the senior executives who decide what goes into newspapers and what runs on the television news. Science specialists do not tend to climb editorial ladders in the same way as political, business or foreign reporters. What this means is that there's minimal geek input to the day-to-day news agenda. Errors in the reporting of science are common, and are tolerated in a way they would not be in business or sport. 'Embryo' and 'foetus', 'vaccine' and 'antiviral' are used as if they were synonyms. Units are converted incorrectly, and basic scientific concepts and statistics are misused.

The editorial idiosyncrasies that will always influence the way news is covered are likely to be those of people without a firm grounding in science. Many a health scare has begun when a tabloid executive fell sick and generalized from his own perceived experience, or concluded that there must be an epidemic. Some editors decide a paper's line on an issue such as MMR with more regard for the way other editors will treat the story than for the evidence behind it. Science specialists are often pressurized to write in a way they would not choose, and find it hard to challenge unreasonably sensationalist framing decided above their pay grade. Good science reporters judge themselves by what they keep out of the news as well as by what they get into it, and good editors notice and appreciate this. But they worry about getting a reputation as the Dr No who is forever knocking things down, when many editors prefer the can-do fixer who makes their wishes come true.

In broadcast news, there is a parallel shortage of scientific awareness and expertise among the senior journalists who anchor the most influential bulletins. This isn't a problem when they introduce specialist reports about science, but it becomes an issue when they conduct interviews themselves.

One of *Today*'s presenters, Justin Webb, has a particularly un-enviable record of allowing his guests to make dubious scientific assertions without challenge. When Ian Plimer, the author of a book criticizing the science of climate change, appeared on *Today* in 2009, he made a string of extraordinary claims. He stated that 'we cannot stop carbon emissions because most of them come from volcanoes', and that 'not one great climate change in the past has actually been driven by carbon dioxide'. Both claims are demonstrably erroneous. Yet Webb (who also refereed the Halvorsen–Finn encounter) allowed them to pass. 'Let Plimer speak, but let his interviewers do some sodding research first,' the environmental commentator George Monbiot wrote afterwards. 'It wouldn't have been hard for Webb to have refuted these claims: Plimer makes them every time he speaks, and they have been debunked again and again.'

In his review of BBC science coverage, Steve Jones singled out another interview by Webb, with a campaigner against GM crops called Kirtana Chandrasekaran, for particular criticism. 'She claimed that GM crops are a risk to health (no convincing evidence, after twenty years of study), that they will contaminate other potato varieties (almost all varieties of commercial potato are grown from tubers rather than seeds; no risk of cross-contamination), that anti-biotic [resistance] genes will spread in the environment and cause health problems (impossible; no genes for medically relevant anti-biotics in the crop itself but they are widespread in the billions of bacteria we eat, with no harm, each day),' Jones pointed out. 'In the face of these claims the programme simply moved on.'

Even Evan Davis, the former BBC economics correspondent who is the *Today* presenter most comfortable with science, is not immune to the odd howler. After the magnitude 9 Tohoku earthquake and tsunami that caused devastation in Japan in March 2011, he repeated a tabloid report of the scientifically ludicrous idea that it could have been triggered by the moon. To his credit, he corrected himself after geeks pointed out his error on Twitter. But it was an illuminating mistake that should never have been allowed to happen in the first place.

The science gap is also visible among the commentariat. Leader-writing teams are replete with experts on politics, foreign affairs and business, but rarely carry a science specialist. The same is true of newspapers' stables of columnists. Only the *Guardian* carries a weekly science opinion column, by Goldacre. Many of the other columnists who engage with science at all are apt to sneer: Simon Jenkins in the *Guardian* likes to repeat the charge that science is too big for its boots. The overall effect is to imply that science lies outside the Premier League of issues, that it is of peripheral importance to politics and society.

That is also the implication of the guest lists for political discussion programmes, such as *Question Time*. The BBC's flagship debate show invites all sorts of public figures to join professional politicians on the weekly panel moderated by David Dimbleby. The extra guests are often columnists, such as Matthew Parris, Polly Toynbee and Kelvin McKenzie. Sometimes they are business leaders, clerics or generals. Historians and novelists are popular choices: Simon Schama was on twice in 2010 and Will Self has appeared four times. Yet scientists hardly ever make the cut.

Just one scientist, Brian Cox, was invited to appear on *Question Time* throughout 2010 – the BBC's 'Year of Science'. Even that was an improvement on 2009, when the programme featured two pop stars (Jarvis Cocker and Will Young) but no scientists at all. In fact, no one you could even loosely call a scientist appeared between Raj Persaud, the psychiatrist, in December 2004, and Cox almost six years later.

This oversight speaks volumes about media attitudes. It is not as if there is a shortage of scientists with the intellectual breadth to contribute to political debate: Sir Paul Nurse, Lord Rees of Ludlow, Dame Nancy Rothwell and Professor Colin Blakemore are just a few big names with interesting perspectives. Their absence is a lingering manifestation of the Two Cultures divide identified by C. P. Snow back in 1959. Science is still seen to sit apart from the serious business of politics and government. It isn't that it's deliberately marginalized; it's that editors and ministers, with limited experience and appreciation of science, fail to consider what its perspective might add to public debate.

Geeking the media

Nobody these days admits to believing everything they read in the papers. Trust in the media is on the wane, and the online revolution means that the leading papers and broadcasters are no longer the only sources of news. But it would be wrong to think that these trends undermine the importance of the mainstream media's coverage of science. Despite all the challenges it faces in the internet age, the media is still the chief channel by which most adults encounter science and make sense of its insights into public health, food, energy and the environment.

The 2011 Public Attitudes to Science report, commissioned from Ipsos-MORI by the UK Government Office for Science, found that the two leading sources of information about science are television (cited by 54 per cent of respondents) and newspapers (32 per cent). Only one in five people named the internet as a major source, and just one in fifty mentioned science blogs. The traditional media remains the prism through which science reaches the public. Television, radio and newspapers also set the tone of political debate. What captures the interest of the *Today* programme, the *Daily Mail* and the *Ten O'Clock News* will also capture the attention of politicians chasing the public's approval. That makes exploiting the mainstream media more adeptly, and changing its approach when we can, an important challenge for geeks.

The media's indifference to science, as in politics, is at once a difficulty and an opportunity. With a few exceptions, newspapers and broadcasters under-represent and misrepresent science not because they are actively hostile, but because the people who run them know no better. That gives us a chance. If we make our case in the right way, there's no reason why we can't greatly improve coverage of science and get critical thinking and respect for evidence a stronger foothold in the fourth estate.

To do this, geeks need to be realistic about how the media works and talk to its various elements in language that journalists can understand. It's no good arguing that the media should see things our way purely because that's the right thing to do, or because newspapers and broadcasters have some sort of duty or responsibility to cover more science in the fashion we would like them to. We can't make the mistake of equating news providers with public information services. They are not the same thing.

Instead, journalists and media executives must be convinced that more responsible and high-profile coverage of science and its value to society is in their interests. It must also be demonstrated that failing to offer this carries a cost. And we need to engage more proactively and productively with the media, so that scientists and those who appreciate their contribution to society can less easily be ignored.

Every media operation is to a greater or lesser extent a business, and geeks are a business opportunity. There are already signs that as we become more and more visible, some broadcasters and newspapers are actively seeking ways of attracting our custom. The success of Goldacre's *Bad Science*, Cox's *Wonders* programmes, the thriving festival scene and Robin Ince's comedy shows has made it clear that there's a bigger market for science covered accessibly but intelligently than had previously been appreciated. The geeks represent a huge audience that has been poorly catered for. In a difficult economic climate for newspapers and broadcasters, geeks offer the rare prospect of growth.

That was part of the logic when *The Times* launched its science supplement in 2009. The goal was to create a product that hadn't really existed before, yet for which there was a significant market. James Harding, the editor, was taken by what he sensed was a growing appetite for science among readers. He often notes that in 2008 there were three stories that reliably increased circulation when they appeared on the front page. One was Barack Obama. The second was the global financial crisis. The third was the Large Hadron Collider. When the paper started *Eureka*, Harding's instincts were borne out. *The Times* sells several thousand extra copies on the days *Eureka* appears, and the magazine is chock-full of high-value advertisers who want to reach its readers. The *Guardian*'s decision to start its science blogging network was based on similar thinking, which has also paid off. In August 2011, the science section of its website attracted almost 4 million page impressions and 2.8 million unique users.

What this shows is that, as consumers of the media, geeks have the power to get more of what they want. By supporting those media outlets that do science well – by buying *The Times* on *Eureka* days, and by reading and commenting on the *Guardian*'s blogs – it's possible to send a very clear message that this approach to science is valued. More importantly, it will make science coverage more profitable. The media listens to the bottom line. If votes are the currency of politics, then sales, hits and ratings are what counts in the media. If editors and the executives who appoint them think that covering science more comprehensively and thoughtfully is going to win them more readers, viewers and listeners, they will do it. It's the language they best understand.

The advent of paywalls at *The Times* and the *New York Times*, annoying as these might be, creates another chance to make our presence felt. Media organizations that ask online readers to pay for access analyse the content that best persuades them to part with cash. By subscribing and reading good science journalism, you are supporting its future: you make its business value clear to those who call the shots.

That's the good cop approach. But we need bad cops too. Just as politicians abuse evidence because they think they can get away with it, so the media misrepresents science, dealing in phoney balance, overblown breakthroughs and hyperbolic scares, because doing so is at once easy and painless. Erroneous but arresting headlines are felt to capture readers and ratings, without much downside. An exaggerated or misleading story such as the *Telegraph*'s account of the 'heart attack jab', or even a cynical pattern of reporting such as the anti-MMR campaigns run by several papers, might attract the odd letter of complaint. Editors might occasionally have to print a response, or even a correction. But the disincentives are rarely sufficient to influence editorial behaviour to any great extent.

The challenge is to make media distortions matter. We need to create effective deterrents to bad science reporting, which will make journalists and their editors think more carefully about the approach they take. Geeks must complain vociferously, and in numbers, when the media gets things wrong. When science is misrepresented we should counter-attack to embarrass the media into doing better. But our criticism should also be as constructive as possible and care should be taken not to alienate potential allies.

Journalistic errors and misrepresentations can be difficult to call to account. Stanley Baldwin's famous maxim that the press enjoys power without responsibility still rings true. Newspapers have few obligations to correct the record, and their editors are often loath to admit fault. Errors or misrepresentations can be reported to the Press Complaints Commission (PCC), but its power to act is limited. It will not, for example, adjudicate on complaints from anyone who is not a direct subject of a disputed story. The BBC Trust, which oversees the national broadcaster's output, and Ofcom, which regulates all broadcasters, have rather more teeth, but they tend to interpret fairness through the prism of balance, which is so unsuitable for science. Perhaps the greater public scrutiny of journalism brought on by the phone-hacking scandal will lead to the introduction of more powerful regulators. But for now, we need to work with what we've got. We can still use the watchdogs to our advantage, to inconvenience and to embarrass the media when it gets things badly wrong.

In January 2010, climate science was attracting unprecedented media scrutiny. The publication of hundreds of stolen emails from the University of East Anglia had led many commentators to allege malpractice by several leading climate scientists, and *New Scientist* magazine had discovered that a claim by the Intergovernmental Panel on Climate Change (IPCC) that Himalayan glaciers could melt away by 2035 was unfounded and unrealistic. The *Sunday Times* followed up with a scoop of its own. It reported that another IPCC assertion, that reduced rainfall could wipe out 40 per cent of the Amazon rainforest, was also wrong, 'based on an unsubstantiated claim by green campaigners who had little scientific expertise'. The story rested on an interview with Simon Lewis, a rainforest researcher at the University of Leeds, and was followed up worldwide. Bloggers who dispute the science of global warming predictably dubbed the incident 'Amazongate'.

Lewis, however, had told the paper that the IPCC's statement was 'poorly written and bizarrely referenced, but basically correct'. While he thought the IPCC should not have referenced the figure to a report from the lobby group WWF, the figure was not wrong. Amazongate was itself a distortion, founded on words twisted to manufacture controversy. When the *Sunday Times* failed to publish his letter setting the record straight, Lewis complained to the PCC. The watchdog sided with the scientist, forcing the paper into an apology.

By taking the time to complain, Lewis did more than defend his reputation and set the record straight. He also created great embarrassment for the newspaper that misrepresented him, which was widely criticized both elsewhere in the mainstream media and in the blogosphere. The *Sunday Times*'s reputation for trustworthiness was damaged. It discovered that distorting science in pursuit of headlines is not always cost-free.

The BBC learned the same lesson in 2006 when it broadcast a three-part series called *Alternative Medicine: the Evidence*. It opened with a remarkable scene. 'In China, a young woman is having open heart surgery,' said the presenter. 'But it's unlike anything you'll see in the West. She's still conscious. Because instead of a general anaesthetic, this twenty-first-century surgical team are using a 2,000-year-old method of controlling pain. Acupuncture.'

That, it transpired, was not the whole story. Acupuncture was not, as the documentary implied, the sole or even the primary method of pain relief. The patient on the operating table had also been given powerful sedatives and painkillers. Simon Singh complained to the BBC Trust, which upheld his criticism. It was a considerable embarrassment for the nation's leading broadcaster.

Both cases demonstrate how complaints can begin to counter media misinformation about science by creating a reputational cost. For this to happen, someone has to complain. Too few people who are traduced by the media stand up for themselves as robustly as Simon Lewis. And still fewer are moved to action when it is science as a whole that is traduced. 'I found it quite shocking that the BBC did that and two million people were misled,' says Singh of his alternative-medicine complaint. 'In a way, what was more shocking was the fact that nobody else complained. I'm not a heart surgeon or anaesthetist, but some people who are must have seen it and been annoyed. There must have been hundreds of people with particular expertise who went in to work the next day moaning and grumbling about it. We need more people like this to speak up.'

Even when complaints such as Singh's and Lewis's are not upheld, they are not without value. A PCC or BBC Trust investigation consumes time and resources, forcing editors to examine their reporting and to investigate what might have gone wrong. Even if there is no admission of error or apology, there is a chance that lessons will be learned.

Informal complaints to editors can also be influential here. Just as a full inbox encourages MPs to think about where they stand on an issue, so editors who receive plenty of letters and emails about a contentious article become more likely to ask themselves why so many readers, viewers or listeners have become upset, and whether it might be worth taking steps to prevent a repeat performance. Senior scientists – the Nobel prize-winners and chief executives of leading research organizations – can play a particularly important role, because their accomplishments, positions and reputations can win them access to the people who make the big decisions. Ministers and captains of industry are not above sharing their views on a paper's coverage with editors or proprietors in the hope of influencing it, but scientific leaders seldom exploit status this way.

There's also mileage in reversing this formula, sending a brief email of praise about reporting you respect and value. If editors know that a writer or reporter's work is going down well, they'll have an incentive to commission more of it.

The blogging revolution

Equally valuable is the forensic deconstruction – practised by
Goldacre and the many bloggers he has inspired – of dodgy
reportage. No journalist likes to have his work derided because of
errors, distortions or misunderstandings, and editors understand that
this does little for the reputation of their titles or programmes. Care
must be taken about attacking individuals – the person whose byline
appears on a bad story is not always responsible for its poor content.
But robust criticism and mockery can be a powerful deterrent.

Since Goldacre began his 'Bad Science' column, the fear of 'being
Goldacred' has made many science and health journalists think twice
about writing stories they know to be rubbish. Just as importantly,
this threat has helped specialist writers to stand up to editorial pressure to write rubbish. Most science hacks want to get their stories
right, but they often need to convince their superiors to drop a dodgy
item, or to avoid a sensationalist or irresponsible approach. The existence of 'Bad Science', and of hundreds of similar blogs, works to
their benefit. They can tell news editors that yes, they could write in
the sensational and inaccurate way they've been asked to – but that
there is every chance that they will be held up to public ridicule as a
result.

Humour and parody can also be helpful, not least because they can
bring criticism of poor reporting to a much wider audience.
Exasperated by the formulaic way in which much science is
reported, especially on the BBC website, the *Guardian* blogger
Martin Robbins wrote an inspired spoof under the headline 'This is
a news website article about a scientific paper'. It continued:

In the standfirst I will make a fairly obvious pun about the subject
matter before posing an inane question I have no intention of really
answering: is this an important scientific finding?

In this paragraph I will state the main claim that the research makes,
making appropriate use of 'scare quotes' to ensure that it's clear that I
have no opinion about this research whatsoever.

If the research is about a potential cure, or a solution to a problem, this paragraph will describe how it will raise hopes for a group of sufferers or victims.

This paragraph elaborates on the claim, adding weasel-words like 'the scientists say' to shift responsibility for establishing the likely truth or accuracy of the research findings on to absolutely anybody else but me, the journalist . . .

'*Basically, this is a brief soundbite,*' the scientist will say, from a department and university that I will give brief credit to. '*The existing science is a bit dodgy, whereas my conclusion seems bang on,*' she or he will continue.

Robbins's post went viral. It was recommended 38,000 times on Facebook, it was retweeted 4,800 times on Twitter, and it received 650,000 page views in just a few days. On the day it was published, it accounted for about 15 per cent of the *Guardian* website's entire traffic. By being funny, Robbins's criticisms reached tens of thousands of people who would never otherwise have read them. It's too early to say whether it will have a systematic effect on the way science is reported, but it was definitely noticed. It certainly made me think.

The success of Goldacre, Robbins and many other science bloggers also demonstrates another way in which geeks can compensate for the shortcomings of the mainstream media. As the internet has removed so many of the historic barriers to publishing, it has given individuals unprecedented opportunities to share their views without the intervention of large media organizations. Blogs and social networking tools such as Twitter allow communities who feel badly served by newspapers, radio and television to create a shadow media of their own.

Few blogs have the reach of a general-readership newspaper or the BBC, and they are often criticized, with some fairness, as echo-chambers in which people who already agree with one another entertain themselves by sounding off. Yet these online communities serve several important wider purposes. First, they provide a forum in which issues that matter to people, but which pass the mainstream media by, can be kept alive. They also create a space in which ideas can grow and participants can acquire the confidence to step outside their comfort zone to campaign and complain. Both these effects were important factors in the Singh libel case, as sympathetic bloggers first persuaded the science writer he had considerable support and then made the mainstream media notice the cause.

A further benefit of blogging is that it can begin to correct the record when the media indulges its bad habit of misrepresenting science and research. In May 2010, the journal *Nature Neuroscience* published a paper about acupuncture from a team at the University of Rochester which purported to demonstrate a painkilling effect on mice. If valid, such results would present strong evidence that this complementary therapy has biological effects that go beyond placebo, and much of the media reported it as such. 'Let's get straight to the point: acupuncture DOES ease pain,' said the *Daily Mail*'s headline. Maiken Nedergaard, the lead scientist, claimed in a press release: 'The new findings add to the scientific heft underlying acupuncture.'

Hardly any of these reports set the study in the context of previous research or properly analysed its flaws. The Rochester paper did not compare the effects of acupuncture with those of sham needles that fail to penetrate the skin, despite many earlier studies, and systematic reviews of those studies, showing that this placebo treatment can induce similar levels of pain relief. Where the mainstream media failed, bloggers such as Ed Yong and David Gorski stepped in. The revealing commentaries they provided on both the paper itself and the press coverage it stimulated are there on the web, easily found by anybody who decides to search for more information.

Institutions' blogs can serve the same purpose. The charity Cancer Research UK, for example, runs an outstanding daily blog that picks apart the science behind news stories about cancer and explains major advances in the field. When newspapers indulge their bad habit of exaggerating preliminary animal research into cures that 'raise hope for millions', it sets the record straight. *Behind the Headlines*, a website run by NHS Choices, does a similar job for a still wider range of health stories.

These examples show how bloggers can provide valuable rejoinders to media failures. They may initially reach fewer people than the original bad journalism, but when worried people take to Google to investigate further, they will find an alternative point of view. Occasionally such posts can go viral in their own right, reaching surprisingly large audiences as they are retweeted, 'liked' on Facebook and re-posted on aggregators such as Digg and De.li.cious. Increasing numbers of people use sites like these as their chief sources of news, relying on people they respect to recommend stories of interest, whether they come from newspapers, broadcasters or blogs. It is sometimes possible to build influence by circumventing the mainstream media. Nowhere is it written that traditional news organizations must forever be the only reliable purveyors of news.

The geeks' approach to blogging is also setting new standards for transparency that the better professional science journalists are starting to follow. Science demands that sources are properly credited and online etiquette requires bloggers to provide hyperlinks whenever possible. This has two important functions: it allows the interested reader to locate further information, and also to check that facts have been reported accurately. This excellent and simple practice is sparingly used by newspapers and broadcasters. It isn't always the fault of journalists: some academic papers issued under embargo, for example, are not yet available on the internet for general reading. But linking should always be encouraged. It can help to build trust, and reporters who show their working are more likely to take the trouble to get things right.

Constructive complaints

Adversarial approaches to bad science reporting are important weapons in the geek arsenal. But they shouldn't always be the first ones we deploy. Start by attacking a journalist in abrasive fashion and he is as likely to become defensive and deaf to criticism as he is to take it on board and change his approach. Plenty of science reporters are broadly sympathetic to the aims of Goldacre and his allies, but none the less brush off their views because they consider them relentlessly negative. Complain constructively, couching criticism as helpful advice, and a surprising number of hacks will listen.

Media professionals are as prone as anybody else to 'cognitive dissonance'. This term from psychology describes the difficulty of holding two pieces of mutually incompatible information in your mind at the same time, and it is easily activated by our own errors. Most journalists like to think they are covering their beat responsibly, thoroughly and accurately. If they are then confronted with a serious error, that information can be difficult to take on board – especially if it's implied that they have made the error because they are slipshod, conniving, venal or stupid. And dissonance is often more easily resolved by rejecting inconvenient facts than by acknowledging them.

To achieve a culture change in media attitudes to science, we must work with the grain of cognitive dissonance, rather than against it. As Abraham Lincoln said, 'If you would win a man to your cause, first convince him that you are his sincere friend.'

I can vouch for the value of Lincoln's aphorism. In 2001, soon after I began covering science for *The Times*, I picked up on a story in *New Scientist* about the forthcoming shutdown of the Large Electron-Positron Collider at the CERN particle physics lab near Geneva. The piece suggested that the Higgs boson – the so-called 'God particle' that is supposed to give matter its mass, and for which the accelerator was searching – was unlikely actually to exist. My story strongly implied that the search for it had been a waste of money. 'The results of these recent experiments have convinced many of the 5,000 scientists working in the field that the whole investigation has been a wild goose chase.'

My interpretation was pretty absurd. While there were indeed physicists who thought the LEP results made it unlikely that the Higgs would be found, that hardly meant that looking for it was a waste of time and money. If anything, the non-discovery of the Higgs would be still more exciting, as it would suggest that an entirely new model of physics is required. Sometimes the null result in science is the most valuable that there is. The existence of the Higgs was in any case far from settled: the view that it doesn't exist was, even in 2001, very much a minority one, and most physicists expect the Large Hadron Collider to find it.

That I recognized my error was largely down to an Oxford University physicist called Tony Weidberg. He would have been well within his rights to rant and rave at my sensationalist misinterpretation, but he took a different tack. In a calm and friendly fashion, he told me that I'd made some mistakes in my copy, but that wasn't altogether rare or surprising in such a difficult and technical field. Perhaps I'd like to come up to Oxford to meet him and his colleagues, to learn a bit more about particle physics and make a few expert contacts I'd be able to call on when I next turned to the subject? It was a textbook example of how to turn media misreporting to your advantage. I don't pretend for a moment that I've never made an error about particle physics since visiting Weidberg's team. But his approach gave me a way of learning from my mistakes for which I remain very grateful.

Acknowledging an error so that it can be learned from becomes much more probable if it's presented as the sort of mistake that a reasonable person could reasonably have made. That applies even when you don't think the error reasonable. Had Weidberg shouted down the phone that my piece on the LEP was crap, I'd like to think that I'd have thought about his criticism and eventually accepted it, but I'm realistic enough to know that I'd probably have reacted defensively and sought to justify myself. We'd both have been worse off as a result.

Criticism is most likely to make a difference if it is framed as friendly advice – at least at first. If a journalist makes a bad mistake, you should tell them about it, but politely. If they react badly – or if they've a track record of misrepresentation – by all means step it up a notch. Write that vituperative blogpost. Complain. Hold them up to ridicule. But it's worth exhausting softer tactics before taking a tougher stance.

These different responses to the media's output, proportionately deployed, could work in concert to create a kind of informal peer-review. Once journalists begin to realize that much of what they publish will be scrutinized in public for errors of fact or inter-pretation, more of them can be expected to ask themselves tough questions before they write. They can learn from the methods of science: that by treating your story as a hypothesis that you must test yourself, you reduce the risk that someone else will do that for you, with potentially embarrassing consequences.

Better journalists, and more responsible media organizations, will realize that all this works ultimately to their benefit, allowing them to produce more informative and trustworthy work that remains interesting and immediate for the present readership, while potentially even attracting new audiences. Even those who take a more cynical approach, calculating that sensation sells while accuracy is of secondary importance, will know that this puts them at risk of ridicule and complaint.

Most science writers, and indeed most journalists of all types, would usually prefer to get things right. Few reporters went into journalism to systematically distort the truth: they want their work to be as accurate as the relevant format can possibly allow. To do that, journalists need the help and cooperation of scientists. Not even specialist writers can be expert in everything: they're as good as their sources. Many welcome constructive criticism. And most would welcome a stronger culture of engagement with the media and the public at large among scientists.

Engaging with the media

Not all the blame for the media's misreporting of science lies with the media. The scientific community has also been slow to appreciate the importance of effective public relations: the value both of responding cooperatively to journalists' inquiries and needs, and of proactively engaging with the media and the public to ensure that their views get a fair hearing. Too many scientists remain unwilling to make themselves available to the media, to promote their work, or to respond quickly and saliently when scientific issues do hit the headlines. The results are to raise the risk of misreporting and to leave a void in public debate that is readily filled by others with inimical goals.

This reluctance is sometimes rooted in a feeling that media and public engagement is a time-consuming chore that distracts from the more important business of doing research. The majority of scientific careers, indeed, will pass without interest from the mainstream media. Yet it is still worthwhile for every scientist to think about how they would explain what they do in straightforward terms should a call from a journalist come in. Most science is funded either directly or indirectly by the public, through the government and charities. That confers a responsibility to be open and accessible. The press, on behalf of the public, has a right to ask you to explain. Its inquiries are also a valuable opportunity to explain what you're researching and why, to share your fascination with a wider audience and enthuse them about the use to which their taxes and donations are being put.

Even when scientists appreciate that communication is sometimes part of the job, many hesitate to engage with the media for fear that they'll be misrepresented or quoted out of context, or that their work will be misunderstood or dumbed down to the point of meaninglessness. As experiences like Simon Lewis's show, these are real risks. They are risks that have to be taken, however, because the alternatives are worse.

At the height of the false controversy over MMR in 2003, Channel Five broadcast a hagiographic dramatization of Andrew Wakefield's campaign against the vaccine. It followed this with a studio debate featuring Wakefield himself, in which many experts on public health, paediatrics, autism and vaccination refused to participate. Principled and understandable as this might have been, it was wholly counterproductive. The event still went ahead, stripped of the voices of the people best placed to challenge Wakefield and explain the facts about MMR to millions of viewers.

When GM crops hit the headlines in the late 1990s, plant scientists were similarly reluctant to engage with the media and make their case. While organic lobbyists and green pressure groups actively courted journalists and fed them stories about the dangers of

GM, too many researchers sat on their hands. They expected the facts and the peer-reviewed literature to speak for themselves, when a more proactive approach was needed.

Fail to return a call, or refuse to comment on a scientific controversy, and very rarely will the journalist go away. More likely, the story will be written without you, possibly with assistance and quotes from people with whom you disagree, and you will have lost your opportunity to influence it. Speak to the reporter, and there's a risk you'll be misinterpreted. Refuse to speak, and the risk begins to approach a certainty. Most journalists who approach scientists are not trying to catch them out; they want to understand what they're writing about more deeply and will welcome your corrections of any misconceptions they might have. They deserve the benefit of the doubt. If you are traduced, then complain loudly, not least so your peers know which reporter to be wary of in the future. But handle media enquiries helpfully and in good faith, and you can usually expect the favour to be repaid.

Not everybody needs to be Brian Cox or Robert Winston. 'Scientists should be good at doing science, just as my bank manager should be good at managing a bank,' says Simon Singh. 'But if they can communicate it exceptionally well, like Steve Jones or Brian Cox, they should be treasured.' Yet a further problem is that very often they are not. It is a sad fact of science communication that many of its best practitioners are not lauded by their colleagues, but sneered at as self-publicists or second-raters. 'You can't move in some physics departments without people slagging off Brian Cox,' says Mark Stevenson, a comedian and science writer. 'It's like that class sketch with John Cleese, Ronnie Barker and Ronnie Corbett. The top research scientists look down on the communicators, who look down on the science journalists. It isn't at all productive.'

While some leading communicators, such as Cox and Jim Al-Khalili, have supportive institutions, others face difficulties. I know of one television presenter who has effectively had to give up a research career because of departmental hostility. Carl Sagan was famously blackballed by the National Academy of Sciences when nominated in 1992, scorned by many members as more celebrity than scientist. I've heard from dozens of young scientists who've been steered away from media work by suspicious lab heads.

This has to stop if science is to get the public profile it deserves. Everybody who works to communicate science, from the PhD student who blogs to the celebrity scientist with her own television show, should be supported and valued, so long as they get things right. We're all on the same side, and geeks need all the media champions we can get. 'For scientists who would be agents of change, communication is not an add-on,' says Nancy Baron, director of science outreach at the marine biology group Compass. 'It is central to their enterprise.'

There are good signs that science is starting to grasp this. Since it was blindsided by media controversies over MMR, GM crops and mobile phones a decade ago, it has begun to get its PR act together. Organizations such as the Science Media Centre and Sense About Science have become effective matchmaking services, pairing up journalists in need of expert comment with scientists who are well placed to provide it. They also run rapid rebuttal and commentary services, collecting print-ready soundbites from relevant experts when a scientific story hits the news. If you work as a scientist and care about communicating science more effectively, you should get yourself on their databases.

These organizations are also becoming adept at helping scientists to use the media as an agent of influence. The successful campaign over the Human Fertilisation and Embryology Bill discussed in Chapter 2 is a case in point. When the government proposed banning the creation of human–animal hybrid embryos for medical research, the scientists involved in this work did not wait for the media to come to them. They seized the agenda, briefing journalists about the threat this posed to British science. They explained the work from first principles, so that it appeared, if not familiar, then at least neither foreign nor frightening. They engaged groups representing patients with diseases such as Parkinson's to explain why the research could benefit them. And they built a coalition comprising most of the country's medical and scientific institutions to call for the legislation to be revised.

The net result was to frame a story that could easily have been presented in the press as an example of unfettered, unethical science that had gone too far in much more favourable terms. From the outset, the media instead pursued the narrative that ministers had proposed over-regulation that would block promising medical research without good reason. Scientists were on the front foot and enlisted the media's assistance to get the result they wanted. The steady diet of embarrassing newspaper stories that they engineered was an important factor in the government's U-turn.

'Between two early press briefings and the final vote in Parliament, the scientists involved had talked to journalists hundreds of times and done scores of media interviews,' recalled Fiona Fox, director of the Science Media Centre, in an Academy of Medical Sciences pamphlet published when the Bill became law. 'Every development from Parliament, the HFEA or Catholic bishops was seized upon as an opportunity to repeat the scientific and human case for pursuing this research, and to correct misrepresentations. No interview – no matter what time of day or night, and no matter how difficult or unpleasant – was turned down by the scientists involved.

'In 1999, during a similar national debate on GM crops, many of the best plant scientists in the country turned away from unsympathetic media that splashed their research on the front pages. As a result, the British public said no to a new technology without ever hearing the case for it from the scientists. Now, ten years later, we can point to a debate on a controversial issue which . . . was frequently dominated by the science.'

This new boldness and self-confidence with which scientists are handling the media bore fruit again in 2010, with two more successful campaigns that demonstrate how geeks can build political influence. As we saw in Chapter 2, the immigration cap's effects on science were significantly watered down following an astute lobbying effort by the Campaign for Science and Engineering and the Royal Society, abetted by a series of reports about its impact in *The Times*. The Comprehensive Spending Review that autumn also brought out the best in geek activism, which ensured that science was spared draconian funding cuts.

It's to that example that we now turn, for while that campaign did as much as could be expected, it has a way to go yet. The case that science is more vital to the economy than most politicians realize is one that still needs to be made.

GEEKONOMICS

Why science matters to the economy

WHEN THE CONSERVATIVE MINISTER WILLIAM WALDEGRAVE CLEARED his desk as Chief Secretary to the Treasury following Labour's landslide election victory of 1997, he left a short note for his successor, Alistair Darling. As the gatekeeper to pots of taxpayers' gold, he advised, Darling would be besieged by special interests arguing that this cut or that was unconscionable. Only two budgets, however, really merited special protection. One was the intelligence agencies. The other was science. 'Investment in science cannot be turned on and off on a political whim,' Waldegrave said as he recalled the letter thirteen years later. 'We must have long-term investment.'

Though Waldegrave wondered whether his message might have been 'confiscated by civil servants on the grounds it might contaminate', Darling read it. And the Labour government made a good fist of heeding its advice: over thirteen years in power, it roughly doubled the budget of the research councils to £6 billion. But by the time it left office in 2010, a different mood had taken hold. With the national deficit pushing £178 billion, Labour's outgoing Chief Secretary, Liam Byrne, left a less helpful note for his successor, David Laws. 'There's no money left,' it said. Major cuts in government spending lay ahead, and science lay squarely in the firing line.

Neither the Conservatives nor the Liberal Democrats, now governing in coalition, had manifesto commitments to guarantee the level of science funding. Only the NHS was to be granted immunity from the deepest spending cuts in a generation, and Adam Afriyie, the Tories' science spokesman in opposition, had made it plain that 'major science budget cuts are inevitable'. Scientists and their supporters began to prepare for the worst. As science is not generally considered a voting issue, arousing popular passions in the same way as the Armed Forces, education or the police, it looked certain to be a prime target for the Treasury. As George Osborne, the Chancellor of the Exchequer, prepared a Comprehensive Spending Review for October, the research councils that distribute public funding for science were asked to draw up plans to save as much as 30 per cent.

The gravity of the situation was underlined on 8 September, when the Cabinet minister responsible for science – Vince Cable, the Business Secretary – delivered his first speech on the subject. He warned scientists that the government's austerity agenda meant that they would have 'to do more with less', and that he was determined to 'screen out mediocrity'. He added, erroneously, that 45 per cent of public funding rewards research that is 'not of an excellent standard'. His implication was clear: science cuts were planned, and senior ministers thought they would be justified.

It was all too much for Jenny Rohn. Angered by Cable's slight against her discipline, the cell biologist at University College London, who is also a published novelist with a beautifully written science blog, called her fellow geeks to arms. 'Sod it,' she blogged. 'Let's march on London! No more Doctor Nice Guy, no more hiding behind our work, no more just taking things lying down like we take everything else in our profession – poor job prospects, poor funding, low pay, poor life-work balance. If they are going to bleed us dry, we might as well try to do something before it's too late. I reckon there are thousands of practising scientists and their allies in the vicinity – let's make some noise. Who's in?'

She got a surprising answer. Positive comments began to pile up at the bottom of Rohn's post and other science bloggers started to link to it approvingly. As her supporters recommended it on Twitter and Facebook, under the hashtag #scienceisvital, the Nature Network, which hosted her blog, crashed under the weight of the unexpected traffic. Then, entirely spontaneously, people who had been inspired by Rohn's blog began to make things happen. Shane

McCracken, a web manager Rohn did not know, created a Facebook page and registered a domain name. Evan Harris, the former Lib Dem MP, and Imran Khan, director of the Campaign for Science and Engineering (CaSE), offered help with a petition, parliamentary lobbying, a media push and a demonstration. Rohn's rallying cry gave voice to the frustrations of thousands of geeks. The Science is Vital campaign was born.

With the spending review announcement scheduled for 20 October, just six weeks away, Rohn's band of geek activists had little time. But just as Simon Singh's allies used social networking to support his libel defence, so online tools allowed Science is Vital to build a popular movement fast. Programmers offered their skills for free to code a petition website and gather the email addresses of likely supporters. McCracken and Della Thomas, a young bio-medical scientist working in London, used Twitter and Google Documents to collate and share the contact details of tens of thousands of scientists, and to recruit volunteers to write to them.

Celebrities such as Brian Cox, Dara O'Briain and Patrick Moore offered enthusiastic backing, as did the journal *Nature* and scientific charities such as Cancer Research UK and the Wellcome Trust. CaSE put up the money for a mass email facility and insurance for a public demonstration. Harris organized a police licence for campaigners to protest outside the Treasury. Julian Huppert, the only scientist in the House of Commons, booked a committee room for activists to lobby their MPs.

The petition grew and grew, and on 9 October, a bright Saturday afternoon, more than 2,000 people assembled for a rally in King Charles Street off Whitehall. 'Hey! Osborne! Leave our geeks alone!' they sang, to the tune of Pink Floyd's 'Another Brick in the Wall'. Speakers including Harris, Ben Goldacre and Professor Colin Blakemore addressed a sea of white coats about the need for continued public funding. The BBC, the *Guardian* and *Metro* covered the novelty of scientists speaking up for their interests. Three days later, hundreds of activists descended on Portcullis House, the office building across the road from the Houses of Parliament, to make their case to MPs. The Science Minister, David Willetts, sent the civil service's Director General of Research, Professor Sir Adrian Smith, to listen and observe. Huppert tabled an Early Day Motion endorsing Science is Vital that was signed by 141 MPs.

On 14 October, several of the campaign's leaders delivered a petition bearing 33,804 signatures to 10 Downing Street, before being invited to meet Willetts. Besides Nobel laureates such as Sir Paul Nurse and Andre Geim, and thousands of working scientists, the document carried the names of housewives and taxi-drivers, musicians and postmen.

Rohn was overwhelmed by the speed and scale of the reaction. 'I was serious and angry, but didn't think anyone would respond,' she says. 'All of a sudden, just a few weeks later, we had 2,500 people outside the Treasury, and I was meeting David Willetts. How did that happen?'

The campaign's impact didn't end there. By the time Osborne got to his feet at the Dispatch Box to deliver the spending review, many of the science cuts that his department had proposed had evaporated. Instead of the 30 per cent reductions that had been pencilled in weeks earlier, the largest element of the science budget was to be frozen at £4.6 billion over the next four years. The 'flat cash' settlement would mean a real-terms cut, the full extent of which would depend on inflation, and capital spending on buildings and equipment would take a much bigger hit. Yet the outcome was far better than most scientists had dared hope for. Rohn's passionate intervention, and the tireless efforts of the people she inspired, had had a tangible impact.

Any doubt as to whether the campaign made a difference was dispelled by the Chancellor's spending review speech. 'Britain is a world leader in scientific research, and that is vital to our future economic success,' he said. His choice of words stood as testament to what geek activism can achieve.

Why science is vital

The breadth and volume of support that Science is Vital mustered made it demonstrably clear to the government that scientists, and people with an affinity for science, were no longer a constituency that could be ignored. That 34,000 people could be persuaded to endorse the petition in just a few weeks sent the message that science was a salient issue to plenty of voters, who would recall the government's decisions about research funding when the next election came around. It demonstrated that ignoring geeks might just carry a cost.

Yet had it involved nothing more than protest, Science is Vital would probably have failed. Plenty of other interest groups took to the streets ahead of the spending review without winning any concessions. The No More Doctor Nice Guy tactics had an impact because they backed up a cogent and consistent argument that the Chancellor could understand: that by cutting research, he would be cutting off his nose to spite his face. What carried the day was geekonomics: the compelling case that public investment in science is vital to economic growth.

By adopting this message, long articulated by the Campaign for Science and Engineering, Science is Vital managed to speak in the mother tongue of the Treasury officials and Conservative politicians it needed to impress. It encouraged its supporters to write to ministers and MPs to impress on them the value of research to UK plc. Science spending, the campaigners argued, is historically proven to be that rare beast: a genuine investment that brings economic returns.

Countries such as Britain aren't rich in natural resources, and high labour costs make our low- and medium-skilled industries uncompetitive. We have to live off our wits. We can do that through financial services, though as the past few years have shown, it's unwise to put too many eggs in that basket. And we can do it through science. The skills provided by a strong science base encourage global companies in key sectors such as pharmaceuticals and aerospace to set up shop in Britain. Both fundamental curiosity-led science and the applied research that translates it into medicine and technology are core drivers of the innovation that economies like the UK's require to thrive and to grow.

As Waldegrave advised Darling, a solid science base is difficult to build up but easily destroyed. Scientific talent is mobile: if stronger financial support is available abroad, many of our best researchers will leave to pursue it. The brain drain would be back. Promising young scientists, too, will not sit around waiting for funding to pick up if their job prospects look grim. They will leave for other countries, or leave science for other professions, choking off the pipeline of talent on which the innovations of the future rely.

By making these arguments, Science is Vital reinforced a message that scientific leaders had been pressing on the Treasury even before the May election that put Osborne into No. 11 Downing Street. In March 2010, the Royal Society had published *The Scientific Century*, a report assessing science's contribution to the economy and the case for public funding. It was prepared by a panel that included former Labour and Conservative science ministers, two Nobel laureates, and senior executives from the pharmaceutical company Pfizer and the IT company Cisco, and its verdict was unambiguous. Cuts to science funding would be a false economy that would risk 'relegation from the scientific premier league'. Science, it declared, is so critical to Britain's economic recovery that ministers should really be spending more, not less, as part of a long-term strategy to build growth.

On the same day, the engineering entrepreneur Sir James Dyson reached a similar conclusion in *Ingenious Britain*, a report commissioned by the Conservative Party. The government needed to invest in science, technology and engineering, and to ensure that innovations that emerge from research can be properly exploited by industry. The Prime Minister's Council on Science and Technology also endorsed the case for sustained investment, particularly given increased spending by rival nations such as the US, France, Germany and China.

As Whitehall departments consulted interested parties ahead of the spending review, science continued to build on this consensus. The Royal Society, often cautious in its discussions with ministers, explained in no uncertain terms that even a 10 per cent cut beyond inflation would start an unprecedented brain drain that would hit Britain's international competitiveness. A bigger cut, of 20 per cent or more, would 'cause key parts of the system to unravel, permanently damaging UK capabilities in key areas'. It headlined this scenario: 'Game over'.

Science's champions also drew heavily on the growing empirical evidence for its economic benefits. A 2010 paper by Jonathan Haskel, Professor of Economics at Imperial College Business School, and Gavin Wallis, who is both a University College London academic and a Treasury official, was particularly influential. It found that public spending on science, through the research councils, has a striking effect on growth.

'Current spend is around £3.5 billion,' Haskel told the Commons Science and Technology Committee. 'This gives around £60 billion additional market sector output. If, to be conservative, one halves this, one gets a contribution of publicly funded research to GDP of £30 billion, which is about 2 per cent of GDP. Put another way, if support for research councils was cut by, say, £1 billion from its current £3 billion, GDP would fall by around £10 billion.'

The economic impact of one science sector was assessed in some detail in 2008 in *Medical Research: What's it worth?*, a report commissioned by the Medical Research Council from the Health Economics Research Group, the Office of Health Economics, and RAND Europe. The return on investment for cardiovascular science was estimated at an astonishing 39 per cent: for every pound spent just once, Britain earned 39p every year. Mental health fared almost as well, with a 37 per cent return for every pound.

The evidence went on and on. Between 2003 and 2007 thirty-one university spin-out companies were floated on stock exchanges, with a combined value of £1.5 billion. In the same period, ten successful spin-out companies were acquired by larger businesses for another £1.9 billion. Put together, they created some £3.4 billion for the UK economy – about as much as the research councils spend in a year. A 2001 study by the Organisation for Economic Co-operation and Development found that increasing public spending on research and development raises productivity throughout the economy. State- and charity-funded research also stimulates research in the private sector: in pharmaceuticals, an extra 1 per cent from taxpayers leads industry to spend an extra 1.7 per cent within eight years.

A further tactical coup was for science to strike an alliance with business. In June 2010, the Campaign for Science and Engineering persuaded senior executives from companies such as Glaxo-SmithKline, Airbus and Syngenta to write to *The Times*. Their significant presence in the UK is no accident, they explained, but a consequence of the strength of its academic science that might be reconsidered if this was weakened.

'The UK's private sector invests £16 billion in research and development and employs 150,000 people,' they said. 'Our companies are careful about where they invest. We value the scientific and engineering talent that flows from the UK's world-class universities and publicly funded research base ... Scientists, engineers, and multinational companies will focus their pursuits in countries with the most favourable intellectual, financial and political environments.'

All this evidence was to provide important ammunition for science's champions within government. David Willetts, Sir Adrian Smith and Sir John Beddington, the chief scientific adviser, distilled the consistent message from these reports and submissions into a compelling case to put before the Chancellor and Prime Minister. 'The argument was: "Why is Rolls-Royce here? Why is Glaxo here?",' said one figure involved in the negotiations. 'This was language the Treasury understood.'

An incomplete victory

The 2010 spending review, like the Simon Singh libel case, illustrates what geeks can achieve when we get our tactics right. The scientific community built a strong case for generous public funding. It framed that case cleverly, in terms calculated to appeal to the decision-makers of the day. It married behind-the-scenes lobbying by senior figures with public protests and media engagement. The result was to ensure that this was an issue the government had to take seriously.

Sources close to the Chancellor have told me that he was always minded to protect science, but in the prevailing climate of austerity he would have struggled to do so without the evidence placed before him. The geeks gave their allies in positions of power the information they needed to secure a much more favourable settlement for science than once seemed plausible.

And yet, for all the success of Science is Vital, government spending on research in Britain is still being cut. Despite a sympathetic Chancellor, armed by an effective campaign with copious evidence of economic benefit, an already small budget was pared back still further.

Resource spending – the funding stream that provides research grants and salaries – is being allowed to erode with inflation, which in 2011 was running at over 4 per cent. A VAT increase from 17.5 per cent to 20 per cent has added further costs. The House of Commons library estimates that total government spending on research in 2014–15 will be about 14 per cent lower in real terms than it was in 2010–11.

The science budget freeze, indeed, was achieved only through an accounting sleight of hand, by redefining exactly what it covers. The Campaign for Science and Engineering has shown that when counted according to the previous government's system, overall spending on science will fall from a planned £5.8 billion in 2010–11 to £5.4 billion in 2014–15.

Capital spending, on buildings, facilities and equipment, is being reduced significantly, by about 46 per cent over four years. The weak pound has put pressure on Britain's subscriptions to international science projects such as CERN, the European Space Agency and the European Southern Observatory, which are priced in stronger Swiss francs or euros. Universities lost much of their state support for teaching science students, and the research budgets of government departments were trimmed.

These cuts might just have been bearable had there been fat in the system that could easily be trimmed. But British science wasn't exactly flush with cash in the first place. In 2009, a year before the cuts, a funding shortfall at the Science and Technology Facilities Council forced it to limit use of the ISIS neutron source, a key instrument for medical and environmental research, to just 120 days a year. It also withdrew financial support from optical and infrared telescopes in the northern hemisphere, potentially denying British astronomers access.

The UK, indeed, has spent a far smaller share of its national wealth on science than most of its leading economic rivals for two decades. While more than 30 per cent of the UK's GDP comes from sectors that rely heavily on science, technology, engineering and mathematics, the government spent just 0.55 per cent of GDP on research and development in these areas in 2007. Of the G7 countries, only Italy spent a smaller share: Germany spent 0.71 per cent, the USA 0.78 per cent and France 0.81 per cent.

Neither is the competition standing still. As Germany and France cut spending to reduce their national deficits, they not only spared science but stepped up its funding. In 2010, France announced an extra €35 billion for research, while Germany pledged to increase spending on science and education by €12 billion by 2013. The European Union's innovation 'scoreboard', published in February 2011, showed the UK well behind its two biggest continental rivals, prompting Máire Geoghegan-Quinn, the European Commissioner for research, to warn Britain that it risks falling behind.

'This is an area, regardless of the cuts which have to be made in other areas of an economy, which needs investment,' she told the *Financial Times*. 'We have China breathing down our necks; we have the US far ahead of us. I think it's disappointing . . . that what is being done in France and Germany is not being replicated in the UK. I'm talking about France and Germany substantially increasing their investment in this whole area, while at the same time that same increase in investment is not happening in the UK.'

Australia increased spending on science by 25 per cent in 2009–10. Canada found an extra C$32 million for its three main research agencies in 2010–11. In the US, President Obama's stimulus package included more than $100 billion for science, and his administration has been clear about its desire to enhance science funding on a more regular basis: the White House's 2011 Budget proposed a 5.9 per cent increase for research in an otherwise flat settlement. That the Republican Congress forced Obama into an Osborne-like compromise, with funding for the National Institutes of Health and the National Science Foundation trimmed slightly, just goes to show that short-sighted politicians are not a species found uniquely in Britain's parliamentary habitat. The House of Representatives, indeed, had originally proposed much more swingeing cuts, including a $900 million raid on the Department of Energy's Office of Science.

Emerging economies, too, have noticed the potential of geeko-nomics. Developed nations such as Britain, France and the US cannot compete with India, China and Brazil over the cost of labour, making highly skilled, high-value-added industries and services based on science and technology essential to their economic futures. India, China and Brazil, however, have not proved willing to sit back and cede this science-based growth to their wealthier rivals. They want a piece of it, and are committing vast sums to science and research accordingly.

Since 1999, China has increased its spending on science by almost 20 per cent every year and now spends $100 billion a year on research: more than thirty times as much as the UK

Since 1999, China has increased its spending on science by almost 20 per cent every year and now spends $100 billion a year on research: more than thirty times as much as the UK. By 2020, it intends to spend 2.5 per cent of its GDP on science, which will mean an annual science budget of some $300 billion. The UK used to have the same percentage target, until the present government cancelled it. China, with engineers and scientists well represented among its political leadership, actually seems serious about reaching it.

A Royal Society report into international science, published in March 2011, noted that China is fast emerging as the next research superpower, which can soon be expected to match and even outstrip the achievements of traditional scientific nations. 'Quantity of input doesn't necessarily result in quality of output, but these investments are starting to yield results,' explains James Wilsdon, the Royal Society's director of policy. 'Since 1981, the number of peer-reviewed papers produced by China has increased 64-fold, and it is well on target to become the leading producer of scientific publications within this decade, perhaps as soon as 2013.'

On current trends, China will begin to file more patent applications than the US, the current world leader, by 2015. 'The impact of China's rise will depend largely on whether we are with them at the technology frontier or onlookers from the sidelines,' the highly respected Institute for Fiscal Studies advised in September 2011. 'We should choose the former.'

A timely reminder that no country has a permanent claim on geekonomics was served in February 2011, when Pfizer, the world's largest research pharmaceutical company, announced the closure of its main UK facility in Sandwich, Kent, with the loss of 2,400 jobs. While the company said the decision was no reflection on UK government policy, John Denham, Labour's Shadow Business Secretary at the time (and a chemistry graduate), has correctly identified that this is less reassuring than it seems.

'It makes the implications worse,' he says. 'A leading company has looked at its worldwide activities and decided they don't need to be in the UK . . . Global companies are not about to leave the UK en masse, but heed the warning. It's no longer enough to be a decent place to do business. Many countries will promise that. We must ensure that the business environment is so good that no leading company can afford not to be here.'

The geeks' campaigning saw off science cuts that would have made it all but impossible for Britain to offer that guarantee. But if this was a victory, it remains far from complete. We have a lot of convincing still to do if science is to fulfil more of its potential to create jobs and growth.

Blind to the consequences

Public spending on science would be worthwhile even if it were a true luxury with no economic payoff. The pursuit of understanding for its own sake is a celebration of human curiosity and supporting it is a legitimate function of the state. What the evidence increasingly shows, though, is that public funding of research is not a luxury at all. It's actually one of the best uses to which taxpayers' money can be put, an investment that is all but guaranteed to garner a sizeable economic return. Science should be getting more public funding, not less.

Why, then, has this message failed to resonate sufficiently with politicians? Why is it that even a chancellor who is broadly sympathetic to science, and who has largely accepted the case that cutting it back is a false economy, has none the less implemented a significant budget squeeze?

The answer lies chiefly in the wider failings of the political classes' understanding and experience of science. As we've seen, only one of the 650 MPs in the UK's House of Commons was a scientist in his previous career, and the US Senate has no scientists or engineers at all. There is a lack of familiarity with the practice of science, and of what it needs to succeed, which blinds politicians to the consequences that their funding decisions will have.

The impact of cuts on scientific careers is a case in point. Much of the science budget is committed to large facilities such as telescopes, DNA-sequencing labs and particle accelerators, which cannot easily be closed to accommodate short-term financial pressures. Mothballing a lab or instrument built with hundreds of millions of pounds of public money is also bad politics. Cuts therefore fall hardest on human resources – particularly on grants and stipends for PhD students and postdoctoral researchers, which are rarely tied up for long periods of time. The Engineering and Physical Sciences Research Council (EPSRC), for instance, responded to the spending review by making plans to cut the number of PhD studentships it supports, potentially by more than a third, from 2,900 to 1,800.

Such short-term savings can have devastating long-term consequences, for they threaten the flow of able people through the system. When promising young researchers are denied the opportunities to develop their skills, their talent will be lost to science, or at least to domestic science, for good. The damage can take decades to unravel: even should a future government decide on a step change in science funding, it will lack the cadre of trained and experienced researchers that will be required to exploit it. 'The internal logic in any bit of the system leads to cutting PhDs,' says James Wilsdon. 'But the aggregate of that would be a lost generation.'

Many politicians go beyond this well-meaning ignorance to sneer at the apparent futility of the work that state-funded scientists do, as risible examples of government waste. When Sarah Palin, the darling of the Tea Party movement, ran for vice-president she high-lighted fruit-fly research as just the sort of meaningless project that taxpayers shouldn't have to fund. Yet as the workhorse of laboratory genetics, the fruit fly has contributed to all manner of medical advances of which one might imagine Palin approves, and the specific project she criticized was spending just $211,000 on investigating a pest that threatens the $85 million California olive oil industry.

John McCain, who chose Palin as his running mate, spent much of 2009 and 2010 attacking science elements of the Obama stimulus package that he thought sounded silly, such as a $71,623 grant for drug research he billed as 'monkeys getting high for science'. Darrell Issa, a Republican congressman, took up the same baton in 2011 with a laundry list of research projects he deemed superfluous and wasteful. These included studies of the effects of video games on mental health in the elderly, and condom use skills among men. Leaving aside the medical and social benefits this work could have, mental illness costs the US at least $200 billion a year, and unplanned pregnancies another $5 billion. Yet research that might address both has been labelled worthless.

Science also struggles in intense competition for public funds because it is perceived as something that can be cut without political penalty, while governments that spend more on research can expect little credit. Even politicians who realize that science funding pays off handsomely in the long run know there is little chance they will be around to make political capital when that happens. The timescales of science work against the sort of quick returns that politicians can use to their advantage. A 2010 report by the Russell Group of leading UK universities, looking at the economic impact of 100 case studies of its institutions' research, found that it took an average of nine years for an initial discovery to yield a product licence and another eight years after that for a spin-out company to turn a profit.

By the time a far-sighted politician who invests in science is vindicated, she is virtually certain no longer to be in the same post and it is quite likely that her political career will be over. By the same token, ministers who swing the axe are rarely around to feel the consequences. Bill Foster, the physicist and former US congressman, points out that cutting science rarely has an immediate deleterious effect. 'The effects are downstream twenty years, while the incentives are to get elected every two years,' he says. 'There's no penalty for doing something that creates long-term economic damage twenty years down the track.'

A further problem derives from a disconnect between how politicians imagine science to drive economic growth and how it really delivers. To ministers, state support for science tends to be a matter of picking winners, of helping out those projects that will lead to lucrative new business opportunities, or that will solve a societal problem such as generating carbon-free energy. In practice, science cannot be directed like that. While investment in science as a whole can be relied upon to bring such rewards in the long term, both the timing and direction of the journey from basic research to technological application to commercial profit are remarkably difficult to predict.

When that journey begins its destination will usually be unknown. Many of the avenues that brilliant researchers choose to pursue will turn out to be blind alleys. Such open-ended research needs to be backed properly, however, even though much of it will go nowhere.

If a few projects are to succeed, many more must be allowed to fail. It is when agile minds follow interesting scientific leads for their own sake that most innovations emerge.

The serendipity of science

When Sir Tim Berners-Lee, a CERN computer scientist, developed the hypertext transfer protocol – the 'http' of web addresses – he wasn't trying to invent a revolutionary form of mass communication that would transform countless businesses and enable the creation of entirely new ones. He was seeking a better way for particle physicists to share data, and his elegant solution happened to give birth to the World Wide Web.

David Payne is not as well known as Berners-Lee, yet the Professor of Photonics at the University of Southampton can be just as fairly called a father of the modern internet. Every time you download a track to your iPod or watch a video on YouTube you are probably making use of his research, which laid the foundations of the fibre-optic cable technology that made broadband possible and brought significant economic benefits in its wake.

In the mid-1980s, Payne's team designed an amplifier to send signals over long distances down fibre-optic cables. It was to have an unexpected extra benefit. 'It turned out you could put multiple "colours" through the fibres, perhaps 1,000 different wavelengths,' he says. 'The upshot was, you could have fibres not only 1,000 times longer, but carrying 1,000 times more information. It's where broadband comes from.'

A critical foundation for a new industry worth many billions had been laid: there are now enough fibre-optic cables to circle the globe 30,000 times. It did not come about because Payne was trying to invent it, nor because he was backed by a big corporation that saw his potential. It happened because the British taxpayer gave his team the time and space it needed by funding excellent scientists to follow their curiosity over many years.

When Southampton's Optoelectronics Research Centre (ORC) made its critical breakthrough in 1987, it had a ten-year grant worth £16 million from the government funding body that is now the EPSRC. It gave Payne and his colleagues latitude to take on long-term projects and to explore exciting new directions as they became clear. Some lines of inquiry went nowhere. Others opened serendipitous technological possibilities and business opportunities that could never have been predicted in advance. Broadband was far from the only economic spin-off: ten companies in the Southampton area, employing a total of 600 people, owe their origins to the ORC. 'We've established their collective turnover at around £200 million a year,' Payne says. 'Our research has more than paid for itself.'

Shankar Balasubramanian has a similar story to tell. When the Cambridge University chemical biologist was growing up as a Merseyside teenager, his ambition was to become a professional footballer and play for Liverpool. While his ultimate career path lacked glamour by comparison, it had far more striking economic consequences. As he played around with nucleic acids in the late 1990s, following his curiosity, Balasubramanian and his colleague David Klenerman hit upon a new method of reading DNA that dramatically speeded up the process. The pair hadn't set out to make a fortune, or even to develop a useful new technology, yet their findings were to prove both important and lucrative.

Solexa, the company they founded to take forward their research, rapidly became one of the success stories of British biotechnology. Solexa sequencing machines are used today by most of the world's leading genetics labs, contributing to a gathering revolution in personalized medicine. In November 2006, Solexa was acquired by Illumina, the world's leading DNA-sequencing company, for $600 million – all money the British economy would never have seen had research councils lacked the resources to support fundamental chemistry research.

More recently, Andre Geim and Kostya Novoselov won the 2010 Nobel Prize for Physics for their discovery of graphene, a new form of carbon that is promising to transform electronics and many other industries. Like most great ideas that emerge from science to generate growth, it was rooted in serendipity, an unforeseen spin-off that emerged when important scientific questions were answered.

Carl Sagan, the master of communicating by thought experiment, encapsulated the theme in a passage from *The Demon-Haunted World*:

'Suppose: You are, by the Grace of God, Victoria, Queen of the United Kingdom of Great Britain and Ireland, and Defender of the Faith in the most prosperous and triumphant age of the British Empire,' he wrote. 'Suppose, in the year 1860, you have a visionary idea, so daring it would have been rejected by Jules Verne's publisher. You want a machine that will carry your voice, as well as moving pictures of the glory of the Empire, into every home in the kingdom. What's more, the sounds and pictures must come not through conduits or wires, but somehow out of the air.'

To address this challenge, the Queen engages the Empire's top scientists in the 'Westminster Project' – a Manhattan Project-style grand initiative to develop just such a revolution in communications. It would, of course, have failed. Yet at exactly the time Sagan asks us to consider, a Scottish geek, following his own curiosity, was drawing up a series of mathematical equations that described electromagnetism. His name was James Clerk Maxwell – and his work was ultimately to deliver television, radio and other aspects of modern telecommunications. It is just that nobody, not even Maxwell, could possibly have predicted this in advance.

With so few politicians having worked in research, few of them instinctively grasp the intricacies of the system's reliance on this sort of serendipity. They are keen to praise Nobel prize-winners and extol the value of scientific solutions to climate change or serious disease. They are less comfortable with the need to support uncertain basic research – essential to the chance of Nobels and breakthroughs, but much of which will fail.

As Simon Frantz, of the Nobel Foundation, puts it: 'When ministers talk about science funding, they almost always announce with pride that the UK has overachieved in the number of Nobel Laureates as a prelude to discussing measures that will ensure that we will never come close to this figure again.'

A common temptation is to insist that funding should be directed so that it supports work that will generate the returns that everyone would like science to have. In 2009, Lord Drayson, then the UK Science Minister, called for funding to be focused on areas of science in which Britain has a 'strategic advantage'. His government also introduced a new funding system for university research, which marks researchers according to the economic or social benefits of their work. Such ideas are well intentioned: it is important that scientists have every incentive to exploit results that have commercial or social potential. But it is simply impossible to predict the impact of scientific research before it is done: if it were known, there would be no need to do the experiments.

More state support for translating good ideas into business opportunities would certainly be welcome. But this cannot be done at the expense of basic research, and the constraints politicians place on science funding makes it something of a zero-sum game. Throw out too much curiosity-led science in favour of applied work and you risk being left with nothing to apply.

> *It is simply impossible to predict the impact of scientific research before it is done: if it were known, there would be no need to do the experiments*

Britain had no strategic advantage in opto-electronics when David Payne began his career: the field had yet to be developed. Shankar Balasubramanian and David Klenerman had no track record of impact when they started the research that founded a multi-million pound company. Berners-Lee would have struggled to show his work had social significance, let alone economic importance. All may have struggled to attract funding in the modern era.

Few researchers are better qualified to boast about impact than Robert Langer, a biological engineer at the Massachusetts Institute of Technology: he holds more than 600 patents and has founded a clutch of successful companies. Yet he is wary of efforts to guide research this way. 'You don't know when you're starting where things are going to lead,' he says. 'I think making things too directed would be concerning. In fact, I would say that generally you get more bang for your buck by funding basic research.'

Changing the game

As is the case for media misreporting of science, political reluctance to give science the financial backing it deserves isn't all the fault of politicians. Scientists and geeks have also failed to make the strongest possible case for a bigger share of the public purse.

With almost all research in Britain, the US and elsewhere funded by the government, scientists are often nervous about biting the hand that feeds them, by asking too publicly or outspokenly for more. They worry that they'll be seen as ungrateful, self-interested or venal, and not without good reason. Global warming conspiracy theorists regard much of climate science as a ruse to wring grants out of governments, and the influential commentator Simon Jenkins has argued that swine flu, BSE and many other health threats have been similarly exaggerated so that scientists can feather their nests. Learned societies and other scientific organizations have thus preferred a cautious, behind-closed-doors approach to lobbying and have rarely voiced public displeasure when decisions fail to go their way. It is as if they have to accept their lot, for fear that complaining might lead politicians to pull the rug from beneath them entirely.

Rolf Heuer, director general of CERN, betrayed this characteristic wariness in a 2011 article for *The Times* which otherwise made an impeccable case for the value of funding basic research. 'Of course, in times of austerity, savings must be made, and if the public purse is shrinking, science must take its share of the load,' he wrote. Much of Britain's research establishment reacted similarly to the 2010 spending review, praising the government for a settlement which, while much better than had been anticipated, still amounted to a substantial cut.

We do need to give credit to politicians who fight to protect science under trying circumstances, as Osborne, Willetts and others did in the 2010 spending review. But we also need to push on afterwards: we can't afford to sit back and be thankful for small cuts when we think they're short-sighted. If unfavourable funding decisions aren't the subject of robust criticism, politicians will not think twice before repeating them.

While we need to strike a balance between realism and fighting our corner, a surfeit of realism can help to perpetuate realities that really need to be challenged. Heuer's statement was written as if it were self-evidently true, when he ought to have been arguing that it isn't. Why must science take its share of the load when it is such a reliable generator of wealth and an engine for escaping times of austerity?

For a profession that is founded on evidence, too, scientists have been slow to collect and examine the evidence for the social and economic value of their own work. It is telling that almost all the studies quoted in this chapter that point towards the long-term economic gains from public science funding have been published during the past ten years. Until recently, there simply hasn't been very much good evidence to back up the case that science funding represents a genuine investment with a reliable rate of return, because collecting it has not been deemed worthwhile.

Scientists who collaborate with business or who start successful companies, too, have often found themselves the object of suspicion from their peers, of whispers that they care more about money than science for its own sake. In Britain, scientists who found companies are asked why; at American universities such as MIT and Stanford, those who don't are asked why not. Just as top communicators of science deserve support from their peers in place of sneers, so too do those who lead the way in commercializing research, even or perhaps especially when doing so brings them personal financial rewards. Scientific entrepreneurs such as Sir Greg Winter, whose company Cambridge Antibody Technology was acquired by AstraZeneca in 2006 for £702 million, and Sir Richard Friend, who founded the successful plastic electronics company Plastic Logic, are personifications of the economic worth of basic research. Their compelling stories can help geeks to tell ours.

As the data presented by Science is Vital (largely assembled by the Campaign for Science and Engineering) show, the scientific community is beginning to grasp the importance of gathering evidence for economic benefits and using it to advance our agenda. But we need to work harder to make the connection in the minds of politicians and the public between the innovative high-tech companies that drive growth and the basic, state-funded science without which few of them would exist or operate in Britain.

'I've made myself unpopular in the university sector for saying this, but there has to be an expectation that what it does will eventually turn into economic benefit, and that that should be demonstrated,' Sir Mark Walport says. 'We need to show why the government should be funding science, and how that funding delivers.'

The 'impact agenda' introduced to British university funding is too blunt an instrument to achieve this, and scientists have been right to criticize it. Individual researchers cannot possibly predict the impact of their work and it is unfair even to expect them to demonstrate benefits from their previous research. There is an element of chance to the insights that will be exploited for social or economic gain, and breakthroughs rely on diffuse contributions from many different teams. But that doesn't remove the responsibility of science to tell a better story about itself and to demonstrate with evidence how it has contributed to economic growth, improved health, environmental protection and quality of life.

As impact assessments seem to be here to stay, scientists need to learn to work with them, to find a way of communicating what their work achieves. This isn't something most researchers are instinctively good at. Anna Grey, the research policy manager at the University of York, made a telling comment when the physics department took part in a pilot impact study. 'Writing about impact is a qualitative narrative and scientists are just not used to writing in this way,' she told *Chemistry World*. 'It's something the physics department identified as a skill they haven't got yet.' It's a skill that every scientist will need to develop.

In 1969, Robert R. Wilson, the first director of the US Fermilab particle accelerator project in Illinois, was called before the Congressional Joint Committee on Atomic Energy. Pressed repeatedly by Senator John Pastore about its potential utility to national security, he replied: 'It has only to do with the respect with which we regard one another, the dignity of men, our love of culture. It has to do with: Are we good painters, good sculptors, great poets? I mean all the things we really venerate in our country and are patriotic about. It has nothing to do directly with defending our country except to make it worth defending.'

He correctly ascertained that the value of satisfying human curiosity about the ways of the universe has a relevance that transcends defence, economics or health. But Wilson's is not the only sound reason for supporting research. It is quite possible to advocate science for its own sake, while also deploying more utilitarian arguments that are more likely to resonate with politicians and the public. Supporting clever people to follow their curiosity, indeed, is the only way we will generate the ideas that lead to innovation and new economic opportunities. Just ask Tim Berners-Lee, David Payne or Shankar Balasubramanian.

'Science is going to have two battles in the years ahead,' says John Denham. 'One is to defend scientific research as of right. The other is to make sure the products of science are used in wider society, especially for the challenge of building a strong economy. Opposing applied science to fundamental science is missing the point.

'I know that people in science, as I do, defend fundamental science in its own right. But the strongest lever we have for scientific research is its short, medium and long-term contribution to a balanced economy.'

We will need these arguments if we're to build the convincing case for better science funding that geeks can and should make. We're also going to have to throw aside caution and reticence about criticizing political decisions over funding and to develop an altogether more ambitious set of demands. It isn't sufficient to be grateful for what we get and bite our tongues instead of asking for more. The idea that science must accept its share of pain when cuts are required, that it will always account for its present small share of government spending, needs decisively to be broken. We can accept the reality of today's funding environment while robustly challenging the logic behind it and arguing for a paradigm shift.

With strong evidence to hand that every pound invested in science will lead to much greater gains in national prosperity, why are we not calling for a step change in science funding? Why aren't we challenging politicians to double or treble the research budget if they're serious about encouraging innovation as a foundation of growth? Thinking big like this could also have tactical advantages. For little financial outlay in the context of overall government spending, a politician who adopted this ambitious agenda could legitimately claim to be rebalancing the national economy with an eye on the future. Politicians are forever in search of vision and narrative; geekonomics can provide these ready-made.

Science is one of the activities that governments should always support, regardless of their political complexion. It contributes to improvements in health, the environment and quality of life. It creates new businesses, while attracting and retaining established ones. These are benefits with cross-party appeal.

From the left, the logic has been accepted by Barack Obama. From the right, it's been endorsed by George Will, the conservative *Washington Post* columnist who has drawn the ire of many scientists for his contrarian opinions about global warming. 'Republicans are rightly determined to be economizers,' Will wrote as budget negotiations between Congress and the White House approached impasse in 2011. 'They must, however, make distinctions. Congressional conservatives can demonstrate that skill by defending research spending that sustains collaboration among complex institutions – corporations' research entities and research universities. Research, including in the biological sciences, that yields epoch-making advances requires time horizons that often are impossible for businesses, with their inescapable attention to quarterly results.'

Investment in science is also worthwhile in all economic conditions. The Institute for Fiscal Studies is generally hawkish about the importance of cutting national deficits, but it acknowledges that this should not mean science spending should suffer. 'Failing to invest sufficiently in science and skills can be short sighted,' it said in a September 2011 report. 'The impact of such spending occurs in the long run, in the form of higher productivity and economic growth. Being able to compete with China in ten years' time requires investment in skills and research today . . . The current economic climate should not prevent investment in our capacity for economic growth in the future.'

All major parties can be persuaded of the merit of this argument. That way, we can depoliticize science funding and make the building of this long-term investment a strategy with consistent and bipartisan political support. This would get around the political short-termism, which often serves science so badly, by creating funding frameworks that last ten or fifteen years and survive changes in government.

Science, says Denham, 'wants consistency and certainty in government policy. That's true for the wider economy as well. Companies looking to invest in this country are not looking for policies that change every five minutes or every five years. If you had a robust ten-year framework for investment in science and innovation that could be supported by both parties, why not, that would be a strength. It's essential that we have science investment that is sufficient to meet the country's economic needs. The starting point is what does the country need to invest.'

The business leaders who wrote to *The Times* ahead of the spending review concurred. 'The UK has an opportunity to drive future economic growth through science and engineering,' they said. 'To do so, the Government needs a clear, strong and long-term strategy for making the UK the most attractive country for companies to conduct research and development. This includes investing in education and public sector research, so that highly skilled graduates and technicians in science, technology, engineering and mathematics are nurtured in the UK.'

A long-term strategy of this sort is what all the geeks who gave such enthusiastic support to Science is Vital should be calling for next, using the same tactics that brought them a limited, if important, victory. The campaign was significant not only because of its results, but also because it wasn't afraid to draw on economic arguments. The scale of the support it amassed in a few short weeks demonstrated the size of the community that cares about this issue. It began to establish that science is a political constituency, with votes to be won or lost.

The challenge now is to push on. With the contact details of tens of thousands of geeks in its database, Science is Vital has a valuable resource that can become a political weapon. High on the list of issues where we should deploy it is the other area of investment demanded by the business leaders. For if we are properly to exploit the potential of geekonomics, the way we educate the next generation is just as important as what we spend on research.

SCIENCE LESSONS

Why science matters to education

THE TIMETABLE AT MONKSEATON SCHOOL ON TYNESIDE SOUNDS LIKE A teenager's dream.

While their peers at neighbouring comprehensives must drag themselves up and into school in time for lessons starting at 9am, its pupils have the luxury of an extra hour in bed. The school day doesn't begin until 10.

They have science to thank for their lie-ins. Adolescents' renowned reluctance to get up early isn't just a symptom of teenage surliness. Research led by Russell Foster, Professor of Circadian Neuroscience at the University of Oxford, has shown that their body clocks actually run several hours behind those of adults, because of differences in the timing hormone melatonin. Nature hasn't made them 'morning people'. This discovery, Foster thought, raised an intriguing possibility. Secondary schools generally insist on an early start, on the logic that their pupils perform best in the mornings. But perhaps this was doing teenagers a disservice.

When Foster first suggested that the typical secondary-school day might begin too early, he was ridiculed. The National Association of Head Teachers described his idea as 'totally inappropriate'. John Dunford, general secretary of the Association of School and College Leaders, said he knew from experience that the opposite was true and that children work best first thing in the morning. Yet Paul Kelley, Monkseaton's headteacher, saw an opportunity to innovate. In 2010, he persuaded the school governors to let him push the first lessons back by an hour.

It was an innovation that appears to be paying off. In August 2011, after the first full school year using the new timetable, Monkseaton's Year 11 pupils recorded the best GCSE results in the school's thirty-nine-year history. The proportion of pupils achieving at least five GCSEs at grades A* to C rose by 19 per cent on the previous year. Results were especially impressive in science and information and communications technology. Persistent absenteeism has also fallen by 27 per cent.

As Kelley is the first to admit, it's possible these improvements would have happened anyway. But at minimum, the change has done no harm. Aside from the recruitment of two new maths teachers, there had been no other substantive changes to the staff or curriculum that could potentially explain the better results. 'Obviously we can't prove the results are down to the timetable, but I've never known for a school to get such big improvements across all the core subjects,' Kelley told *The Times*.

Monkseaton's experience is a great example of the way science can inform educational experiments. As research reveals more about the workings of the human brain and body, and about the biological idiosyncrasies of children and teenagers, some of these insights are likely to have implications for the way we teach and learn. We aren't going to find out unless headteachers like Kelley are encouraged to try out new hypotheses. Yet as the attitudes of the teaching unions demonstrate, much of the educational establishment seems curiously uninterested in what science might have to offer.

It isn't only the study of circadian rhythms that might plausibly help schools to get better results. It is now known from psychology that there are three good ways of predicting how well a pre-school child will fare later in life. One is the child's IQ. Another is socio-economic background. The third involves marshmallows.

In the late 1960s, the Stanford University psychologist Walter Mischel devised a simple test for three-year-olds at campus nursery. Each was given a treat, such as a marshmallow or cookie, and a choice. The child could eat it when Mischel left the room, or leave it for fifteen minutes until he came back. If she waited, she'd then get a second treat as a reward. The children who were able to delay their gratification had better exam results when they left school almost two decades later.

These results were recently confirmed and extended by a major study that followed the development of more than 1,000 people in New Zealand between birth and the age of thirty-two. The research, led by Terrie Moffitt and Avshalom Caspi, of Duke University and King's College London, showed that children as young as three who had poor self-control skills were more likely to grow up with a wide range of health and social problems. Besides educational attainment, self-control affected children's later risk of obesity, high blood pressure and sexually transmitted infections, and social issues such as drug abuse, criminal convictions, financial troubles and unwanted pregnancies.

There is growing evidence that these critical self-control skills can be taught. A systematic review of thirty-four studies, led by Alex Piquero, of Florida State University, found in 2010 that role-playing exercises, rewarding self-control and several other interventions held promise for teaching delayed gratification. Another review, led by Adele Diamond, of the University of British Columbia, has found that computer training, martial arts and yoga can have a beneficial effect.

If psychology is well placed to improve teaching now, neuroscience promises to do likewise in the near future. Its potential to unlock the mechanisms by which the brain learns was highlighted in 2011 by a Royal Society working party chaired by Professor Uta Frith, of the Institute of Cognitive Neuroscience at University College London. 'Education is about enhancing learning, and neuroscience is about understanding the mental processes involved in learning,' it concluded. 'This common ground suggests a future in which educational practice can be transformed by science, just as medical practice was transformed by science about a century ago.'

As understanding of cognition and the brain advances, it is suggesting that there may be 'critical periods' for certain aspects of learning, such as foreign-language skills, and that there may be important differences in the way boys and girls learn, especially around puberty. The fine motor control needed to manipulate a pencil or pen develops fully only around the age of five – which may have important implications for teaching handwriting, which commonly begins at a younger age.

Some of these discoveries are more ready to inform educational practice than others. The evidence for self-control training is probably strong enough for it to play a much greater role today. Neuroscience, on the other hand, has little to say directly just yet. But it is hard to believe that a better understanding of the infant, child and adolescent brain will prove irrelevant to teaching. Just as improved understanding of immunology and genetics has enabled smart, science-led drug design, so a better grasp of how young brains grow and learn should allow the design of teaching techniques that go with the grain of nature rather than against it.

The usefulness of these scientific insights, however, can't just be assumed. It isn't good enough simply to adopt new teaching techniques because they appear consistent with the findings of neuroscience or psychology, any more than it is to teach a certain way because that's how it's always been done. Though many drugs are now developed to target biological pathways that are known to malfunction in disease, we do not accept they are safe and effective just because they are cleverly designed and scientifically plausible. We know that even brilliant ideas, put together with the best possible understanding of the current science, can be and often are completely wrong. So we assess new medicines exhaustively, in randomized controlled trials (RCTs), before they are given widely to patients.

Kelley remarked, correctly, of his timetable experiment that it's impossible to be certain whether it caused his pupils' better results or whether the improvements were mere coincidence. But there is a way of finding out. We can put educational hypotheses like this to the test. If that's to happen, though, education needs to develop a geekier mindset.

Teaching to the test

When people are asked what they consider to be the most important issue on the political agenda, education generally ranks second only to health. It is understood and accepted by British politicians of all parties that providing universal access to first-class medical care and providing excellent state schools that offer opportunity to all are among the core functions of government. Yet when it comes to evidence, these two central policy concerns are held to entirely different standards.

In healthcare, we expect drugs and medical procedures to be assiduously tested by the most rigorous appropriate methods before they are licensed for general use and before the state agrees to fund them. For the most part, this means assessment by randomized controlled trials – the most reliable method yet devised for determining whether or not a particular intervention really works.

The RCT is commonly described as a kind of 'gold standard' for medical research because it seeks systematically to minimize potential bias through a series of simple safeguards. It must have a control group, so an intervention is tested against another intervention – an inert placebo, or an existing drug. This ensures that the influence of the treatment itself and not the act of treatment is measured. Patients must also be allocated at random to each group to ensure that a researcher can't prejudice the results consciously or unconsciously by, for example, picking sicker patients as controls.

RCTs are not without drawbacks. Critics have argued that it is unethical to experiment on human subjects this way, or to give patients a treatment that may be less than optimal. This type of research can be expensive and time-consuming, and there is often debate over the applicability of conclusions when the tight parameters of the trial are varied in the real world. But in medicine at least, RCTs are accepted as essential because they generate more dependable knowledge than any other approach. The alternative to experimenting this way is not to avoid experiments. It is to conduct uncontrolled, non-random experiments which raise much greater ethical problems because they do not generate useful data.

RCTs have become commonplace in medicine because we accept that the evidence they produce is of paramount importance when life and health are at stake. Life chances and health are also profoundly affected by the educational opportunities that children receive, but we demand a different order of evidence for the teaching techniques that our schools employ.

There is no good reason why RCTs should not be used routinely to evaluate different approaches to teaching children to read, or to educate children with special needs. Yet they are not a standard part of the educational landscape. By the same token, too many new initiatives pass without formal evaluation, and plenty are un-encumbered even by the need to collect useful data. Education, which could easily be a bastion of evidence-based policy, is replete with policy-based evidence.

The question of how best to teach children to read is among the most controversial issues in education. For many years the favoured approach was to teach children to recognize whole words. Then in the 1990s and 2000s, more and more teachers began to be won over by an alternative strategy, synthetic phonics, by which children learn to match sounds to letters and groups of letters. A seven-year study in Clackmannanshire, Scotland, suggested that children taught using phonics were three and a half years ahead of their peers in reading by the time they finished primary school. More and more parents began to demand it.

In 2006, an expert inquiry led by Sir Jim Rose, a former director of inspection at the Office for Standards in Education, found an 'overwhelming case' for phonics. Ruth Kelly, the Education Secretary, duly mandated its use in primary schools the following year. Michael Gove, the current Education Secretary, is presently extending provision.

In parallel with the Rose review, the government asked a team led by Carole Torgerson, then of the University of York, to examine the evidence for phonics. Her report noted that while there were indications of promise, these came principally from the US school system and that effect sizes were small. The highly cited Clackmannanshire study was the only substantial evidence from a UK context, and it had not been randomized. A second Clackmannanshire study had used random allocation, but had other weaknesses: the sample size was very small, and children in the intervention and control groups had been taught by the same teacher, creating clear potential for bias.

In short, robust evidence was lacking. If phonics was to be more widely adopted, Torgerson advised, the government should roll it out gradually, with the first areas to benefit being chosen at random. That way, it would be possible to determine whether or not it really worked.

'That was never done,' Torgerson says. 'It just became policy. We had a real opportunity to do a randomized study that might have settled this issue, and it was missed. As a result, we still don't know whether or not phonics works as the main teaching strategy for all children. Some of the recent evaluation work has demonstrated synthetic phonics may not be having the impact that was hoped for. If we'd done the randomized trial we would have known before the policy went national.'

This isn't the only case in which opportunities to conduct RCTs have been spurned. In 2005, the government began what was supposed to be a three-year pilot study of the Every Child a Reader programme. It offered a one-to-one intervention called Reading Recovery to the lowest-achieving 5 per cent of children in the first two years of compulsory primary education.

As with synthetic phonics, however, the pilot wasn't a randomized trial. Worse, it was then deemed such a success after a single year that the programme was implemented nationally, before it had even finished. When the Commons Science and Technology Committee challenged Carole Willis, chief scientific adviser at the Department for Children, Schools and Families, about this, she said a randomized trial would have been too expensive and that it was too difficult to persuade schools to take part. Yet neither problem prevented the department from running an RCT of a parallel mathematics intervention, Every Child Counts. Her arguments, the committee found, 'do not stand up to scrutiny . . . we conclude that a randomized controlled trial of Reading Recovery was both feasible and necessary.'

As this book was being written, a new government was missing another opportunity. In June 2010, Michael Gove announced a radical plan to convert the 200 weakest primary schools in England into academies, free from local authority control, to improve standards. Whether it will work is anybody's guess, and it will probably remain so. It would have been simple, Torgerson says, to turn this initiative into a proper trial, by selecting 100 of the schools at random to get academy status immediately, with the others following the next year. That first year would have allowed researchers to compare outcomes, to establish whether or not there was a positive effect.

The experimental timetable devised by Paul Kelley and Russell Foster could also be evaluated this way. Schools that are otherwise similar could be selected at random to start lessons at 9am or 10am. We'd know after a while whether the innovation raises standards – or whether John Dunford's confidence that the old methods are best is well founded.

The politics of evidence

Politicians and civil servants tend to overlook the importance of educational RCTs for several reasons. Their general lack of familiarity with the methods of science means that they have little appreciation of their value and that there are different kinds and qualities of research available, some of which provide considerably richer and more useful evidence than others. They also see education as a political problem to which the solutions are principally political and ideological, not evidence-based and technical. School standards are best improved by insisting on discipline, selection and traditional teaching methods if you're right wing, or by fighting elitism and deprivation if you're on the left. The evidence matters only in so far as it supports these preconceived positions.

This explains why politicians are so reluctant to evaluate the impact of their policies. As we saw in Chapter 3, there are considerable disincentives to measuring the effects of your policies too rigorously. 'When you do a carefully designed, well-conducted, well-reported trial of a policy, it leaves you with a result you can't ignore, because there are no alternative explanations,' says Torgerson. 'That's why we do RCTs. The problem for the politicians is if they've already implemented the policy, and the trial shows it's not effective. What should the Government do?'

The answer is often to avoid collecting the data in the first place. In 2004, the government set a national target of bringing all school science laboratories up to a satisfactory standard by 2005–06 and to a good or excellent standard by 2010. Yet a National Audit Office report published in 2010 found that this target was not backed up by routine collection of official data so that ministers could be held to it. The report identified more than 470 different initiatives in operation in 2004 aimed at improving participation and achievement in science and maths. 'Some two thirds had no evaluation or none was planned,' it found.

RCTs are also held back by a perception among politicians, education authorities, teachers and parents that they are unethical. Carole Willis summed up the argument when she told the Science and Technology Committee: 'Some local authorities or schools perceive it as unfair that some of their pupils will be getting some sort of intervention that others are not.'

This objection, however, holds no water. A well-designed RCT does not involve one group of children 'getting some sort of inter-vention that others are not'. Groups of socially matched children will be given different interventions of unknown effectiveness and the outcomes are compared. Alternatively, a policy is rolled out at random so that everybody benefits in the end, but the later adopters can serve as a control group for study. It isn't a question of one group of willing participants getting no help at all. The ethics are no different to those involved in medical trials; indeed, where participants' health is involved, the ethical issues raised are, if any-thing, greater.

Every new intervention in children's education is to some extent an experiment. What marks such experiments out as ethical is when they are properly conducted so that we can learn from them – and that generally involves randomization and controls. 'It's not un-ethical to do experiments in education,' says Sir Mark Walport. 'It is unethical not to. Many of the educational policies that are put into action are experimental as it is. They are just experiments without controls. When I raised this in Whitehall, I got the response: "You don't possibly expect to compare educational outcomes like this, do you?" But that missed the point. The point is that you do this when you don't know. The scientific method, which is to ask the question, is almost the antithesis of the political method, which is to say I'll tell you the solution.'

A more rigorous approach to trials in education would leave it much less vulnerable than it is to unscientific novelties. As there is no framework of evidential standards that are required before new techniques are adopted by schools, many have spent taxpayers' money on pseudoscientific gimmickry. Brain Gym, 'an educational, movement based programme aimed at enabling children and adults to reach their potential', is a prime example. This proprietary set of exercises is sold to schools as a way of improving pupils' learning capacity. Yet as the company that sells them admits, the evidence that they achieve anything is largely anecdotal. Many of their purported effects also contravene known physiology and science.

'It's not unethical to do experiments in education. It is unethical not to.' –
Sir Mark Walport, Director of the Wellcome Trust

The 'Brain Buttons' exercise involves rubbing on either side of your breast bone. This is said to promote the brain's capacity for 'sending messages from the right brain hemisphere to the left side of the body, and vice versa; receiving increased oxygen; stimulation of the carotid artery for increased blood supply to the brain; an increased flow of electromagnetic energy.' As Ben Goldacre explained in *Bad Science*: 'Children can be disgusting, and often they can develop extraordinary talents, but I'm yet to meet any child who can stimulate his carotid arteries inside his ribcage. That's probably going to need the sharp scissors that only mummy can use.'

Countering claptrap like this, incidentally, is another important job for geeks. In 2008, Sense About Science put together a briefing on Brain Gym in which neuroscientists and physiologists debunked many of its claims. 'We got loads of calls from junior teachers thanking us,' says Tracey Brown, the charity's director. 'They weren't happy at all about being asked to do this, and the document gave them a reason to go in and question it with their heads.'

Changing the culture

Education's reticence towards evidence-based teaching isn't by and large the fault of individual teachers. Rather, it's the fault of policymakers who don't understand or don't want to understand its importance and value, and of a professional culture that doesn't attach sufficient value to research.

In this respect, it again has much to learn from medicine. In late adolescence, typically between the ages of seventeen and twenty-four, most of us will acquire a set of wisdom teeth. These back molars, however, have a nasty habit of growing at an odd angle so they press on other teeth. In the dental jargon, they become impacted. Until recently, a case of impacted wisdom teeth meant one thing: even if you had no pain or infection, you would have them whipped out. The procedure was the most commonly performed operation in the UK. Then, in the mid-1990s, Jonathan Shepherd began to look at the evidence.

After reviewing thousands of prophylactic operations, the professor of maxillofacial surgery at Cardiff University came to a startling conclusion. There was no evidence that removing impacted teeth in the absence of symptoms led to better outcomes – and the risks of infection and nerve damage far outweighed the benefits. His conclusions have transformed dental surgery: the NHS no longer funds prophylactic removal, and the number of operations has plummeted. Surgeons have stopped performing many unnecessary and costly procedures because one of their number thought to ask whether the technique was really worthwhile.

Shepherd was led to evaluate prophylactic wisdom-tooth surgery in part because of the value that medicine places on evidence. But there was another reason for his curiosity as well. Like many doctors, he is a practitioner-academic. Shepherd is a consultant surgeon who treats patients, but also a researcher whose job description includes designing trials of new techniques and reviewing the evidence base for old ones.

These dual professional responsibilities contributed directly to his work on wisdom teeth. In his clinical practice, Shepherd noticed that the outcomes of his patients who were referred for prophylactic surgery did not seem noticeably better than those who were not, so his training and professional curiosity led him to investigate using the most rigorous tools available to him. 'It is an extremely powerful and productive way to run a profession,' Shepherd says. 'Anecdotal experience from clinical practice drives you to generate hypotheses, which you then have the research skills and experience to put to the test.'

The practitioner-academic culture has contributed to medical progress time and again: in Shepherd's own field of surgery, he cites the advent of day-case operations in place of in-patient procedures, and the treatment of coronary heart disease with angioplasty instead of bypasses, as two advances that stemmed from the insights of working clinicians being put to the test. It not only means that many doctors have the know-how to do research, but the familiarity with research that it breeds also makes doctors more receptive to using the findings of research in their daily practice.

A cadre of clinician-scientists in teaching hospitals means that research and everyday medicine are closely integrated, while even doctors who do no research are consistently encouraged to keep abreast of the latest evidence through continuing professional development, and even to critique it. Almost every hospital department or GP surgery has what's known as a journal club – a weekly or monthly session in which one doctor presents an interesting research paper and others deconstruct its strengths and weaknesses. It's a very healthy way of making sure everyone knows how much evidence matters.

Few schools have a journal club. And practitioner-academics such as Shepherd are rare in education. There are educational researchers, and there are teachers, but there are very few teachers who are both. 'The day you go into education research is usually the day you stop teaching children regularly,' says Shepherd, who has made it his mission to show other professions what they might gain by emulating the medical approach. 'They are two separate worlds. This makes researchers more likely to ask the wrong questions, and teachers less likely to act on the results.'

Alom Shaha, a physics teacher and film-maker, agrees. 'To me it seems obvious: the stuff I do in class feeds ideas about how we should be teaching, and we need to go away and research those ideas properly, to find out whether they're right. Too few teachers think like this, or combine their teaching with research. I know lots of people who used to teach then went off into research. They don't do both.'

Shaha has recently taken a six-month sabbatical to take up a 'teacher fellowship' with the Nuffield Foundation – one of the few schemes that allow working teachers to become practitioner-academics. He is analysing research into the value of practical science experiments so that he can develop evidence-based resources for teachers. He's also involved in an online journal club for science teachers, which convenes fortnightly on Twitter to debate a paper. Such opportunities should be made much more widely available to teachers of all subjects so they can use their unparalleled knowledge of the contemporary classroom to contribute properly to research and ensure its findings are then shared widely among their colleagues.

Teachers need accessible academic literature that keeps them abreast of the latest research, on the model of the *British Medical Journal*, and they should be expected to keep up with developments and discuss them with their peers. They need time to learn and to ask questions, as well as time to teach.

With a little more of it, they might also be better placed to deliver another aspect of education that has let science down for too long.

School science lessons have traditionally concentrated on teaching the findings of science. They have been less successful at communicating how science is actually done: the nature of the scientific method and its importance as an unparalleled device for interrogating nature and society. While children are taught about what scientists think, and the knowledge that their work has yielded, they learn less about how scientists think and why their approach to understanding is so valuable.

How science works

Susan Blackmore, the psychologist and author, often runs classes in critical thinking for nurses and other health professionals to teach them the basics of evidence-based medicine and good experimental design. She finds that her lessons always work best if she starts with a thought experiment. 'Suppose we find that the more often people consult astrologers or psychics, the longer they live,' she might begin. 'Why might that be true?'

Her students start to suggest ideas. Most know that astrologers and psychics can't actually predict the future, but perhaps they might make people feel better about themselves, with benefits for their health. Maybe healthier people, or people who have already lived long lives, are more likely to believe in the paranormal. After a while, someone will eventually come up with the most probable explanation: women go to psychics and astrologers more often than men, and also live longer. Knowing your horoscope might be associated with longer life expectancy, but it isn't responsible for it: both are the result of something else.

Blackmore's exercise is a marvellous tool for explaining an important principle of science: that correlation is not causation. If you want to determine cause and effect, it isn't enough just to observe that certain behaviours or experiences are reliably associated with certain outcomes. You have to eliminate alternative explanations that could account for your observations as well.

'Once you greet any new correlation with "correlation is not causation" your imagination is let loose,' says Blackmore. 'You find yourself thinking: OK, if A doesn't cause B, could B cause A? Could something else cause them both or could they both be the same thing even though they don't appear to be? What's going on? Can I imagine other possibilities? Could I test them? Could I find out which is true? Then you can be critical of the science stories you hear. Then you are thinking like a scientist.'

Teaching this sort of insight, which teenagers are perfectly capable of grasping, ought to be a primary goal of science education. It matters that people grow up familiar with photosynthesis and the inverse square law, which explain the world around us and have no less capacity to inspire wonder than Homer or Shakespeare. It matters more, though, that people grow up to understand how to distinguish between assertions that have been tested according to the methods of science and granted provisional approval, and those that are founded on faith or anecdote. That too many children leave even great schools without grasping this contributes to the limited affinity for science we have encountered in politics, government and the media. It is a failure that must be addressed if society is to make more of the potential applications of science in public life.

As Carl Sagan asked in *The Demon-Haunted World*: 'If we teach only the findings and products of science – no matter how useful and inspiring they may be – without communicating its critical method, how can the average person possibly distinguish science from pseudoscience?'

My own educational experience bears this out. I was fortunate enough to go to a leading independent school, where I benefited from what can only be described as outstanding teaching. In my science classes, I remember learning about Newton's laws of motion, the periodic table and genes and chromosomes well enough for three As at GCSE. But even this top-of-the-range school science never quite communicated how science is not just about the stuff we know, but just as importantly about the stuff we don't. I gave up science when I reached the sixth form and went on to read history at university. It was only when an editor at *The Times* serendipitously asked me to cover the science brief that I really began to appreciate the rigour and beauty of science, to value it as the ultimate satisfier of a curious mind.

To their credit, the guardians of the science curriculum in the UK have begun to appreciate the need to embed the guiding principles of science more deeply in classes. In 2006, a learning strand known as How Science Works (HSW) was introduced to the national curriculum for secondary schools, forming a key element of both GCSE and A-level courses. It seeks to introduce pupils to the ideas and practices that lie behind science as well as the knowledge with which science furnishes the world, and also to communicate some of the issues that scientific discoveries raise for society. It tackles questions of how scientists know what they claim to know, the limits of that knowledge, and the balance between technology's risks and benefits.

Science is not just about the stuff we know, but just as importantly about the stuff we don't

HSW ought to be a breath of fresh air in the science syllabus, a valuable addition that helps to set science in its proper social context and provides schoolchildren with the scientific know-how that will help them to become better-informed citizens. It should give people the discriminatory skills they need to distinguish science from pseudo-science, and help them to know what questions they need to ask about new scientific ideas to evaluate their worth and implications. It should also demonstrate the relevance of science and its methods to all manner of contemporary challenges – one of the themes of this book. Its introduction, however, was controversial, with critics arguing that it has replaced hard facts with half-baked philosophical musing, that it was a symptom of dumbing-down.

Many teachers and scientists feared that spending more time on the philosophy of science and less on the facts would leave students poorly prepared to pursue science and engineering at A-level and university, and that it would encourage pupils to develop opinions about scientific issues before they have a sound understanding of the scientific concepts involved. Some even argued that it could encourage a relativistic, post-modern view of science as just one of many 'ways of knowing', which could become a Trojan horse for creationism and other forms of pseudoscience. Some pupils, too, have complained that the courses are taught in a formulaic fashion, that they lack proper content and that the answers expected of them in exams require common sense rather than informed learning.

These criticisms, unfortunately, carry some weight, because HSW has been implemented poorly. Examination boards have often set questions that lack appropriate rigour; Alom Shaha describes some as 'ambiguous, subjective and downright unscientific'. 'Many of my students have complained that their science exam questions are now more like English comprehension exercises,' he says, and those are some of the better ones.

A 2009 GCSE biology question set by the AQA board invited candidates to match four 'theories of how new species of plants and animals have developed' – Darwinism, Lamarckism, creationism and intelligent design – to short descriptions. Leaving aside that only Darwinism is a theory – a theory, in science, requires robust supporting evidence – several of the descriptions were manifestly inaccurate. Intelligent design was purported to be based on 'the complicated way in which cells work', when it is in fact founded on a profound misunderstanding of the complicated way in which cells work.

There's nothing wrong with addressing creationism and intelligent design in science classes if it is done as a way of exposing the difference between ideas that are developed from and supported by evidence and those that originate from faith alone. But the curriculum and examinations cannot be allowed to offer them any kind of equivalence with actual science.

HSW has also suffered because the science teachers who are expected to teach it have not been given the professional training that is required if it is to enrich their lessons. The strand is, to all intents and purposes, an aspect of philosophy of science, and that is not a discipline in which most science teachers have much experience at all. Few have studied it themselves, either at school or at university, and it wasn't part of their teacher training.

'How Science Works was a much needed change in the science curriculum,' says Shaha. 'The problem is that, as philosophers of science sometimes put it, most scientists know as little about the philosophy of science as fish know about water, and the same applies to science teachers. They just put it in front of teachers and said,

"Teach it." They're being asked to teach How Science Works when they haven't really thought about how science works themselves.'

Many science teachers are also being held back for another reason over which they have little control. They are too often required to teach subjects in which they lack proper expertise. And the current government's policies are threatening to make that situation worse.

Third-class policy

Few people have done more to popularize sums than Carol Vorderman. In twenty-six years as the glamorous numbers whiz on *Countdown*, Channel 4's flagship words-and-numbers show, she made maths fun for thousands of children and inspired many of them to take their homework a little more seriously.

When Michael Gove was Shadow Education Secretary, he sought to tap Vorderman's expertise in popularizing mathematics by appointing her to lead a teaching taskforce. Yet when he graduated to government in 2010, he decided he did not want people like her to teach. For all her talent and passion for maths, Vorderman lacks something that Gove has deemed more important. When she read engineering at Cambridge University, she achieved only a third-class degree.

As part of a drive to make the teaching profession 'brazenly elitist', the Education Secretary decreed that only graduates with a lower second or better would henceforth be eligible for public funding to train as teachers. Your ability to enthuse children might match Vorderman's, but if you also match her degree class you can now teach in a state school only if you pay for your qualifications yourself. While his goal of raising the esteem in which teaching is held as a profession and the quality of individual teachers is well intentioned, Gove's initiative threatens to damage educational standards in maths and science that already leave a lot to be desired.

What Gove overlooked was that in several key science subjects – physics, chemistry and maths – the real problem facing British schools is not the standard of teachers' degrees. It is a severe shortage of teachers with any specialist background at all.

One in four maths teachers in state secondary schools is not a specialist in the subject, and 16 per cent of lessons are not taught by maths graduates. Only 22 per cent of science teachers hold a chemistry degree, and 14 per cent a physics degree, of any class. The Institute of Physics estimates that more than 4,000 specialist physics teachers are needed if every pupil is to have the opportunity to learn the subject from a specialist; Gove's department has accepted this and has set an annual recruitment target of 925 physics graduates each year. This is an ambitious goal: as only about 3,000 physics students graduate annually, the Department for Education is hoping to attract one in three of them. It needs every candidate it can get.

Science graduates with good degrees, however, have many opportunities other than teaching. They are in such demand from academia, industry and finance that they are spoilt for choice; teaching might be able to demand the best humanities graduates, but it does not have that luxury in science. One in four physics teachers who began training in 2008, one in five maths teachers and one in six chemistry teachers had a third. Take them out of the equation and the serious teacher shortage is more likely to get worse than better – even with generous new bursaries of up to £20,000 available for maths, physics and chemistry graduates who meet the 2:2 requirement.

Gove's well-meaning initiative will mean in practice that key science subjects will be likely to be taught by teachers who might have excellent degrees, but in a different field. And this, according to the National Audit Office, will not be good for standards. 'Teaching is of better quality where teachers hold qualifications in the subjects they teach,' it declared in a 2010 report on science education.

A government taskforce chaired by Sir Mark Walport, director of the Wellcome Trust, reached a similar conclusion. 'Our overarching recommendation is that specialist teachers and their subjects need to come to the fore in the delivery of science, technology, engineering and mathematics (STEM) education,' it found. 'This requires teachers with a higher education qualification in their subject area, trained how to teach their subject as well as in general educational skills.'

A strong consensus has been ignored. As a result, much of a generation is going to be taught the laws of thermodynamics by biology graduates who may not have studied physics themselves since the age of fifteen. These ersatz physicists may do their best, but they can hardly be expected to inspire a love of a subject they are uncomfortable with themselves. 'I fear I'll be meeting more and more people who will tell me they hated physics at school – and it won't be their fault,' says Alom Shaha, who does have a degree in the subject. 'And it won't be the fault of their physics teachers either, because, strictly speaking, these people won't ever have had an actual physics teacher.'

A scientific nation?

This sort of misguided political meddling matters because Britain, as we saw in the last chapter, is a nation whose economic future depends heavily on science, engineering and technology. That, in turn, means that growth and prosperity require a school system that provides a steady stream of young people with the scientific skills that are needed if geekonomics is to flourish.

School science is also the prism through which children will see the discipline as they grow up. Get it right and even those pupils who never go on to study science at university, or to pursue a scientific career, will go through life with a sound appreciation of its wonder, its contributions to the modern world and the power of its experimental methods. It should at once be preparing the scientists of the future and making more informed citizens of the rest of us.

Britain's education system too often achieves neither. Some of the best evidence for this has emerged from a series of 'state of the nation' assessments of science education by the Royal Society, the most recent of which, for the 16–19 age group, was published in February 2011. While it found that the overall uptake of science at A-level and its equivalents is increasing, the proportion of young people who study a science subject to a reasonably advanced standard still remains disconcertingly low.

In the 2008–9 school year, about 299,000 teenagers in England and Wales were studying for at least one A-level or equivalent qualification. Of these, only slightly more than one in four (28 per cent in England and 27 per cent in Wales) was taking at least one science subject. Things are a little better in Northern Ireland, where 37 per cent study a science.

The situation is particularly acute for certain subjects. Some 17 per cent of schools offering A-levels in England, 13 per cent in Wales and 43 per cent in Northern Ireland failed to enter a single candidate in physics, while a similar number entered just one or two. As it's hard to believe that nobody in these schools had the necessary aptitude or potential, bright pupils appear to be missing out on the chance to explore a core science subject because they lack encouragement and opportunity.

A 2010 analysis of twenty-four countries by the Nuffield Foundation found that England, Wales and Northern Ireland were the only ones where fewer than one in five pupils studied maths in upper-secondary education (13 per cent for England, 11 per cent for Wales and 15 per cent for Northern Ireland). In Germany, more than nine out of ten pupils study at least some maths at this stage, and all Japanese, Korean and Swedish pupils do.

These low participation rates mean the pool from which universities recruit science students is too small. As a result, fewer than 10,000 British students graduate in physics, chemistry and maths from British universities each year. This output, the Royal Society found, 'leads to a deficit of STEM graduates available to enter employment in commerce and industry'. More than four in ten employers report difficulties recruiting graduates with appropriate STEM skills. It also contributes to intense employment competition for science graduates, which in turn makes it difficult for schools to find specialist teachers. This creates a vicious circle: uninspiring teaching from non-specialists doing their best leads to fewer pupils studying science, so fewer graduates with the right qualifications are available to put things right.

To Dame Athene Donald, Professor of Physics at the University of Cambridge, who chairs the Royal Society's education committee, this 'could be hugely damaging to the prospects of both the individual student and our nation as a whole. There is a need for action on an unprecedented scale to address this problem if we are to ensure our economic competitiveness in a world that is increasingly dependent on science and technology.'

What gives rise to this damaging shortfall, and what can we do about it? One answer comes from the nation that is missing from the paragraphs above: Scotland. North of the border, about 50 per cent of all pupils taking Scottish Highers study a science subject. These account for more than one in three of all teenagers in the relevant age bracket.

One reason for this is the broader character of the exam system: while only the most able pupils generally sit more than three A-levels, it is common in Scotland to sit five Highers, or a combination of Highers and lower-level Intermediates. What this means is that there is less need for pupils to specialize at a young age. Those with a parallel interest in science and the humanities find it simpler to keep up both aspects of their education. The Royal Society has called for a similar qualification that is broader in scope to be introduced across the UK, so that many more sixth-formers continue with science. The International Baccalaureate, which requires candidates to take at least one science course, would be one option. Another would be to adopt a proposal by the Campaign for Science and Engineering to award extra university admissions points for science subjects.

Careers advice is another issue. This can play a critical role in capturing and retaining schoolchildren's interest in science, but the Department for Education admitted to the National Audit Office (NAO) that such provision was 'patchy': only 18 per cent of the 1,274 schoolchildren and young people it surveyed were satisfied with the STEM careers guidance they had received. This leaves far too many pupils unaware of the great demand for people with scientific qualifications and the full range of job opportunities that these can present, and removes a key motivation for studying science.

A further factor is the availability of 'triple science' – the chance to take independent GCSEs in physics, chemistry and biology in place of a double science award or a single combined sciences GCSE. The NAO found that this is strongly linked to uptake of science courses at higher levels, and to higher grades. Yet while provision is increasing, triple science is not an option in almost half of all state secondary schools.

Separate physics, chemistry and biology courses are more commonly available in schools in rich parts of the country, and in the independent sector, but the benefits aren't simply an artefact of this. 'Recent research shows that, compared with other pupils, pupils from more deprived backgrounds achieve relatively larger improvements in their future A-level science and maths outcomes when offered triple science at GCSE than when offered only combined science,' the NAO found. Triple science is least widely available in the deprived areas where it could most enhance children's educational opportunities. An incentive to introduce it has also gone. To call themselves 'specialist science schools', and thus to qualify for extra funding, schools previously had to offer separate science GCSEs. No longer.

Stepping back a stage further, there is also an issue in primary schools. In his introduction to the Royal Society's 2010 report into 5–14 education, Lord Rees of Ludlow, then its president, noted: 'Children are innately curious about the natural world. But, year after year, large proportions are "turned off" science and mathematics by the time they reach secondary school, with little prospect of that interest being rekindled. Inevitably, those who are most likely to suffer are the under-privileged.'

One of the chief reasons for this is a severe shortage of scientific skills among primary-school teachers. There are 193,000 qualified teachers in England's 17,000 primary schools; the Royal Society found that only 5,989 of them had a background in science and 3,903 had a degree in maths. As some of these will be concentrated in the same schools, as many as three quarters of English primary schools lack any member of staff with the academic background to teach science from a position of knowledge and – just as importantly – to support less-qualified colleagues who have to tackle the subject.

It's easy to argue that primary-school science doesn't need to be taught by a specialist, as the content is by definition basic. But there is abundant evidence that science qualifications make a significant difference to teaching quality. A 1997 study by Wynne Harlen and Colin Holroyd found that teachers short on confidence tend to use 'safe' teaching methods that rely heavily on work books, underplay questioning and discussion, and avoid experiments that use unfamiliar equipment. The Royal Society noted further that: 'Teachers lacking a robust understanding of the subject matter and how to teach it are more likely to be influenced by the content of tests than those who have the confidence to know that effective teaching will achieve good results without teaching to the tests.' It recommended that the number of science specialists in primary schools be tripled. Every school should really have one. Even if a single teacher can't be expected to cover all seven primary-school age groups, his or her presence would help other teachers to approach the subject with greater assurance.

That all these assessments, from both inside and outside the government, point to the importance of specialist science teachers underlines the folly of Michael Gove's misconceived elitism. The Royal Society predicts that the 2:2 requirement will 'slash the supply of people into science and mathematics teaching by about 50 per cent. Raising the bar in this way would merely exacerbate problems and mean that some potentially good teachers with poorer degrees would be precluded from joining the profession. We cannot afford to do this.'

The Campaign for Science and Engineering (CaSE), and learned societies such as the Royal Society of Chemistry and the Institute of Physics, have joined the Royal Society in making this point to the minister. Such vocal lobbying, as with so many areas of public policy in which the interests of science are passed by, is going to be important if this bad decision is to be reversed. One key will be to provide the minister with an exit strategy, a way in which he can limit the most damaging effects of his mistake without having to execute a full-blown U-turn. In this case, that could be an exemption to the 2:2 requirement for shortage subjects: maths, physics and chemistry would certainly make the list. And if Gove or whoever succeeds him does eventually back down, let's praise him for listening to the evidence and then changing his mind.

Those who can, teach

Political action isn't the only way in which geeks can take active steps to improve science education. We can also play a participatory role. A growing number of science graduates who have decided that a career in research is not for them are now seeking to become professional science communicators – working in journalism, broadcasting, online media, museum work or public relations. The supply of talented people seeking good jobs in these fields, however, comfortably exceeds demand. Might some of them be persuaded to teach instead?

Alom Shaha, who has worked in both science communication and teaching, is in no doubt as to which career option has the bigger and more lasting impact on society. 'If you want to communicate science, if you really want to make people love and understand science, you can have a far more real and measurable impact as a science teacher,' he says. 'If you work at the Science Museum, you might get to entertain a kid for an hour. As a teacher, you get a captive audience of 200 kids several times a week. A lot of brilliant creative people are wasting their time in science communication, when they could be teaching.'

Not everyone with a passion for enthusing others about science, of course, has the aptitude or motivation to teach. There is still plenty, though, that such geeks can do for education on a practical level. Our schools don't just need more science teachers, they also need more active support from people who know and understand science, who can offer work experience and careers advice, or serve as mentors and sounding boards for teachers. As the Royal Society noted of primary schools, many teachers fail to fascinate their pupils about science not for want of trying, but because they lack the background and knowledge to approach the subject with confidence. That is something that working scientists, and non-scientists with a good understanding of it, can help to change.

The foraging behaviour of the buff-tailed bumblebee, *Bombus terrestris*, is just the sort of material you'd expect to be examined in the pages of *Biology Letters*, a peer-reviewed journal published by the Royal Society. As you'd also expect of an august organ, the study on the subject it published in December 2010 was described in an accompanying commentary as 'a significant piece of research' with results that 'provide convincing evidence that bees can transpose between learned colour, pattern and spatial cues when encountering changes in a coloured scene'.

This, though, was no ordinary paper, as another of its conclusions suggests. 'We also discovered that science is cool and fun because you get to do stuff that no one has ever done before,' its thirty contributing authors wrote. It had been conceived, conducted and written up by a group of twenty-seven pupils at Blackawton Primary School in Devon, aged between eight and ten at the time of their experiments. The remaining three names on the paper belonged to their teacher, Dave Strudwick; their scientific mentor, Beau Lotto, of University College London; and the lead author, one 'Blackawton, PS'.

The publication – the first of its kind – was a fitting conclusion to a groundbreaking collaboration between a working scientist and a schoolteacher that gave the Blackawton pupils an unparalleled opportunity to learn about science by actually doing it for themselves. It came about when Lotto, a neuroscientist whose son Misha attended the school, offered to devise a class science project that would allow the children to conduct a genuinely original piece of research. Instead of simply doing traditional school experiments with right and wrong answers, they'd ask an open-ended question and work out an appropriate series of tests with which to answer it.

Lotto and Strudwick chose bumblebees as the subject and provided the children with a Plexiglas 'bee arena' filled with fake blue or yellow flowers. But the pupils themselves took it from there. 'We came up with lots of questions, but the one we decided to look at was whether bees could learn to use the spatial relationships between colours to figure out which flowers had sugar water in them and which had salt water in them,' they recorded in their paper.

By conducting a string of experiments with the artificial flowers, the children established how the bees used colour to find the desirable flowers filled with sugar water. They progressively ruled out alternative hypotheses to work out how the bees learned to find food. 'We discovered that bumblebees can use a combination of colour and spatial relationships in deciding which colour of flower to forage from,' the children concluded.

It is quite an achievement for children of this age to produce a piece of scientific work of sufficient originality and quality to warrant publication in a prestigious journal. The project's bigger achievement, though, was to introduce the children to an aspect of science that is in many ways its essence, yet which is often poorly communicated in schools. What the Blackawton pupils learned is that science isn't just about looking up the solutions to questions that other people have already answered successfully. It's also the best method that humanity has for solving problems that don't yet have an answer. It is a tool for satisfying our curiosity and answering questions. As Strudwick put it: 'Science shouldn't be seen as something that is detached from the real world – it's just a certain way of looking at things. This project represents a completely different way of working and learning that I hope will be taken up by other schools and in other subjects.'

Too few schoolchildren at any level get the opportunity to devise and run original experiments in this way. But as Lotto's involvement shows, it isn't difficult for ordinary schools – Blackawton is what Alastair Campbell might have called a bog-standard state primary – to offer this sort of experience if they're given the right support. Individual scientists, or even geeks without scientific training, can make these opportunities possible by offering their skills and enthusiasm to schools. Their contributions can give teachers the confidence to create innovative and inspiring science lessons.

The Blackawton project isn't the only one that's worthy of emulation. At the Simon Langton Grammar School in Canterbury, sixth-form students are every year given the opportunity to take part in real scientific research – a recent project involved helping to clone a gene associated with multiple sclerosis. Research Councils UK and the Wellcome Trust run a Researchers in Residence scheme, which allows PhD students and postdoctoral scientists to spend time in secondary schools. David Spiegelhalter, Winton Professor of the Public Understanding of Risk at the University of Cambridge, has developed a fantastic range of aids for teaching schoolchildren about risk and statistics; one uses the National Lottery and the Premiership football table to explain concepts such as chance and regression to the mean. Oxford's Centre for Evidence Based Medicine provides outstanding resources for adding the subject to science lessons.

Then there's *I'm a Scientist, Get me out of here!*, an X-Factor-style online competition between working scientists, in which schoolchildren play the parts of Simon Cowell and Cheryl Cole. Pupils get to quiz participating scientists about their work and their careers, then classes vote for a winner. It provides a terrific opportunity for children to engage in a dialogue with working scientists, to find out how and why they do their research, and to learn about the process of science and its social significance.

All these initiatives demonstrate how people who care about science, but who aren't trained to teach it, can contribute tellingly to science education. Our input is needed. If we can inspire a generation with an affinity for science and a respect for evidence, we might just manage to build a society that demands these qualities in its leaders.

That, in turn, might bring better science to another arena of public policy that could do with it: the criminal justice system.

DOING SCIENCE JUSTICE

Why science matters in court

ON THE AFTERNOON OF 24 JANUARY 2008, DAVID CHENERY-WICKENS walked into a London police station. His wife Diane, a successful make-up artist who had won an Emmy award for her work on the movie *Arabian Nights*, had travelled up by train with him that morning for an appointment at the BBC, or so he claimed. When she failed to meet him afterwards, Chenery-Wickens reported her missing.

On 15 May, a dog-walker reported a strong smell in a wood near the couple's home in East Sussex. It was Diane's body, so badly decomposed that she had to be identified from dental records and the cause of death was impossible to ascertain. The next day, her husband, who worked as a psychic healer, was arrested and charged with her murder.

There was plenty of circumstantial evidence against Chenery-Wickens. CCTV revealed that Diane had not accompanied him on the train to London. He had had a string of affairs, which his wife appeared to have discovered: she had rung mysterious numbers listed on their phone bill on the day police believe she was killed. He had sold some of his wife's jewellery on the day before he reported her missing. But in the absence of forensic evidence about the manner of her death, a conviction would be difficult to secure.

It was the Forensic Science Service (FSS), a state-run company that conducts crime-scene investigations, that cracked the case.

When the victim's watch and ring were discovered in her home after her death, FSS scientists identified tiny bloodstains on the items. High-sensitivity DNA analysis showed the blood to be Diane's and the spatter pattern showed that both must have been removed while she was still bleeding. The jewellery had been removed after her death and brought back to her home – probably in an attempt to hinder identification, as Diane wore both items constantly. An empty vial had been found near Diane's body, leading the defence to claim she had committed suicide, but an FSS toxicologist discredited this: there was no trace of any drug in her system.

The wealth of forensic evidence was enough for the jury. On 2 March 2009, Chenery-Wickens was convicted of his wife's murder and sentenced to life imprisonment with a minimum term of eighteen years.

The murder is one of thousands of crimes which might never have been solved without the FSS, which conducts about 60 per cent of forensic investigations in the UK. Its multi-disciplinary expertise in analysing trace DNA, blood-spatter patterns and toxicology combined to create an unassailable case against Chenery-Wickens. You might expect it to be the kind of scientific success story that would be a cause for national celebration. Yet by the time this book is published, the FSS will be no more. In December 2010, the government announced that the service is to be closed by March 2012.

The FSS, according to James Brokenshire, the Home Office minister responsible, was losing £2 million a month, a sum that was no longer sustainable. Private-sector competitors and police in-house labs, the government argued, were perfectly capable of taking over its functions. The state-run FSS, indeed, was distorting the market and needed to be privatized. 'There is no justification for the uncertainty and costs of trying to restructure and retain the business,' the Home Office declared.

It was a decision that shocked and dismayed forensic scientists. While commercial operators can certainly handle routine forensic investigations, they have little incentive to retain the breadth of expertise that is needed in rare and complex cases such as the murder of Diane Chenery-Wickens. Neither are they likely to invest in the costly training and research that have always been part of the FSS's mission.

'Purely commercial suppliers in such a competitive structure are forced to reduce costs to levels that cannot support the type of research, innovation and attention to case-specific needs that has characterised the commitment to service shown by the FSS,' noted an international group of experts in a letter to *The Times*. 'We are completely convinced that certain key services to the justice system cannot be provided on a purely commercial basis, as there will always be special demands for analytical methods particular to the circumstances of the case that cannot be economically offered, but have to be kept in the inventory of methods available in the quest for establishing the scientific facts in the service of justice.'

The conclusions of a House of Commons Science and Technology Committee inquiry were equally damning. Ministers and officials failed to consult the Crown Prosecution Service or the Criminal Cases Review Commission, nor did they seek the views of the companies they expect to take on the FSS's functions. The chief scientific advisers to the government and the Home Office were also kept in the dark, and were informed of the decision to close the service only once it had been made.

The Home Office had exaggerated the FSS's monthly losses and glossed over the fact that increasing use of police labs that work to lower standards was largely responsible for its financial troubles. The company's extensive forensic science archive could be broken up, interfering with research and with the capacity to conduct cold-case reviews. The closure had also been rushed, so that police authorities were forced to sign contracts with private forensic services without having time for 'due diligence' on their credentials.

A botched ministerial decision, undertaken without proper scientific advice, has threatened the quality of the forensic evidence available to the police and the courts. 'We do need to ask a very serious question, whether the current proposals carry the risk of damaging the system of justice,' said Andrew Miller, the committee's chairman. 'I think there's a serious risk, the way things have been structured. We are going to end up with a mess and the loser will be justice.'

Trial and error

The criminal justice system should by rights have much in common with science. Both are in the business of evaluating and testing evidence, and weighing and minimizing uncertainty. Juries are supposed to find suspects guilty only when a case is proven beyond reasonable doubt; a similarly demanding standard might be said to apply to scientific hypotheses that acquire the status of theories. Science, too, has much to contribute to justice. The forensic evidence it can provide is increasingly important to convicting the guilty and exonerating the innocent.

Yet neither the police officers, judges and lawyers who administer justice, nor the politicians who frame the laws and sentencing regimes they work with, are reliably intelligent consumers of science. They are apt both to place excessive weight on scientific evidence that would not survive proper scrutiny, and to cast aside more robust research that challenges prejudices and preconceptions.

The result can be appalling miscarriages of justice. In 1996, Sally Clark, a solicitor from Cheshire, gave birth to her first child, Christopher. He died suddenly when he was just eleven weeks old. Less than a year later, she had another son, Harry; he was found dead in his cot at eight weeks. A month after Harry's death, both Clark and her husband, Stephen, were arrested on suspicion of murder, and Clark was charged.

When her case came to trial, she had the misfortune to encounter Professor Sir Roy Meadow, then regarded as one of Britain's most eminent paediatricians. Meadow had made his name by characterizing Munchausen Syndrome By Proxy – a pattern of child abuse in which parents harm their children to seek attention or sympathy. Often called as an expert witness when parents claimed cot death but abuse was suspected, he became known for formulating 'Meadow's Law'. 'Unless proven otherwise,' it ran, 'one cot death is tragic, two is suspicious and three is murder.'

In court, Meadow declared that the odds against natural cot death in a non-smoking, well-off family like Clark's were 8,543:1. The odds against two such deaths occurring by chance, he argued, were the square of this figure, or 73 million to 1. The jury returned a 10 to 2 verdict of guilty.

Meadow, however, had committed a litany of statistical sins in compiling this eye-catching figure. As the Royal Statistical Society noted in a public comment on the case, squaring the odds of a one-off death was entirely inappropriate for sudden infant death syndrome (SIDS). 'It would only be valid if SIDS cases arose independently within families, an assumption that would need to be justified empirically. Not only was no such empirical justification provided in the case, but there are very strong a priori reasons for supposing that the assumption will be false. There may well be unknown genetic or environmental factors that predispose families to SIDS, so that a second case within the family becomes much more likely.'

The mistake was particularly culpable as the prosecution was aware that Harry Clark had a bacterial infection, raising the probability of death from natural causes. This evidence was not disclosed to the defence.

Meadow's evidence was misleading for two other reasons. First, the court never heard that if one in 73 million families suffers a double cot death from natural causes, there will be such an event in a country like Britain, with a population of 60 million, every year or two. A single case of this sort should not necessarily be regarded as unusual or suspicious. The odds against winning the Lottery are 14 million to 1, but somebody wins the jackpot most weeks.

His testimony also fell foul of what is known as the 'prosecutor's fallacy'. Let's assume for a moment that double cot deaths are pretty rare, though not quite as rare as Meadow presumed. Perhaps they occur in one in 500,000 families. That doesn't mean that the odds against innocence in such circumstances are one in 500,000. You have to compare the relative likelihood of such a rare event against other competing explanations. Double murders of infants by a parent are also very rare – it's not unreasonable to think these might also occur in one family in 500,000. If that's the case, there was a 50–50 chance that Clark's children were murdered or died naturally.

These figures are illustrative: the chances of a double cot death in the same family are higher than one in 500,000 because of the genetic and environmental factors alluded to by the RSS. But they demonstrate how Sally Clark was failed by the criminal justice system. It allowed the jury to be swayed by misleading evidence that was given the authority of science.

Clark was eventually cleared at her second appeal, in 2003, but died from alcoholism four years later. The courts' susceptibility to bad science may well have claimed her life.

Unchallenged statistical error is not the only source of misleading court science. In television shows such as *Crime Scene Investigation* DNA evidence is usually presented as more or less conclusive. Yet the results of genetic fingerprinting often carry considerable uncertainty that is not always communicated to judge and jury.

In January 2006, Mary Jackson (not her real name) was attacked in a parking lot in Sacramento, California, and forced to perform oral sex on her assailant. Charles Richard Smith was arrested and swabs from his penis identified DNA from a second person mixed with his own. This DNA must have belonged to someone with whom he'd had sexual contact, and if it matched Ms Jackson it would be a strong indication of guilt.

At Smith's trial, a forensic scientist testified that the chances that the sample did not come from Jackson were just 1 in 95,000. Smith was convicted and jailed for twenty-five years.

Genetic evidence, however, can be analysed in multiple ways. The analyst who provided the 1 in 95,000 figure was convinced that he saw reliable 'peaks', indicating matches, at most of the thirteen places in the genome where American forensic scientists compare DNA. His supervisor, whose evidence was also presented, thought fewer of these matches were reliable, and so put the probability that the DNA wasn't Jackson's at 1 in 47.

A subsequent review of the case used a different technique, based on a computer algorithm, to compare the likelihood of the different interpretations of the evidence advanced by the prosecution and the defence. This suggested that this pattern of evidence was only twice as likely if the DNA was Jackson's than if it belonged to someone else.

A recent investigation by the *New Scientist* journalist Linda Geddes used Smith's case to show how apparently convincing DNA evidence isn't necessarily as clear-cut as it might appear. She also identified further subjectivity problems that emerge when different forensic scientists interpret mixtures of DNA from more than one person or from poor-quality samples.

Working with Itiel Dror, of University College London, and Greg Hampikian, of Boise State University in Idaho, Geddes obtained DNA samples from a gang rape in Georgia, for which Kerry Robinson was convicted of being a participant. The samples, together with Robinson's DNA, were sent to seventeen experts for analysis. Importantly, the samples were sent blind, without any contextual information to indicate whether Robinson was a suspect.

At Robinson's trial, two forensic scientists testified that Robinson could not be ruled out as one of the men whose DNA was present in the crime-scene sample. That conclusion was endorsed by only one of *New Scientist*'s seventeen experts: twelve said his DNA was not in the mixture, and four said the evidence was inconclusive.

Such cases do not mean that DNA evidence is worthless – far from it. But they do illustrate the sensitivity with which such evidence must be handled, but often isn't. Private-sector laboratories often refuse to publish details of their techniques, claiming commercial confidentiality, making it difficult for third parties to evaluate them. Geddes's findings particularly question the wisdom of allowing police forces to commission forensic analysis from in-house labs, which are usually unaccredited and lack the arm's-length independence of providers like the condemned FSS.

'There's pressure from prosecutors and investigators to get a result,' Geddes says. 'They're going to their own police labs and saying: "We think we've arrested our guy. Here's his DNA. Can you find him in the sample from the crime scene?" What you really want is something more like a clinical trial, where the analysis is blinded.

'You want one person, with no affiliation to the case, to look at the swab of DNA from the crime scene. He'd make a decision, ideally according to published guidelines, about where the reliable peaks are. Then a second person, also blinded, would look at the suspect's profile and compare it to the results from the crime scene sample. That person should be looking at control samples too, as in a line-up.

'DNA is just as susceptible to biases and subjectivity as other forensic techniques like fingerprinting, bite mark analysis and gunshot residues. The public and jurors and judges and lawyers understand that about these other techniques, and treat them accordingly. DNA still has a shroud of infallibility around it.'

Reasonable doubt

Other branches of law enforcement have proved still less adept at using science, suffering from a surprising inability to evaluate what it can and cannot do. The UK Border Agency, a division of the Home Office, has the difficult and politically sensitive job of judging claims for asylum made by refugees arriving in Britain. In 2009, it decided it might be able to use science to distinguish genuine asylum-seekers fleeing war or persecution from economic migrants falsely claiming asylum to get into the country.

It alighted on two promising techniques. Genetic ancestry testing can provide reasonably robust information about a person's long-term geographical provenance by looking at certain sequences of their DNA. The other procedure, isotope analysis, examines the hair and nails for chemical signatures that are peculiar to certain parts of the world, which reach the body through the diet. Under the Human Provenance Project, the Border Agency sought to use both techniques to identify Kenyans attempting to pass themselves off as refugees from war-torn Somalia.

Geneticists and isotope experts were scornful. The Border Agency, they said, had made the elementary mistake of confusing genetic ancestry and ethnicity with nationality. As David Balding, a population geneticist at Imperial College London, told John Travis of *Science* magazine: 'Genes don't respect national borders, as many legitimate citizens are migrants or direct descendants of migrants, and many national borders split ethnic groups.'

Isotope specialists were equally sceptical, pointing out that such analysis at best shows where someone has been living for the past year of their life. 'It worries me as a scientist that actual people's lives are being influenced based on these methods,' said Jane Evans, of the National Environment Research Council Isotope Geosciences Laboratory in Nottingham.

These weren't minority views, but consensus scientific opinions. Such advice could easily have been obtained by any geneticist or isotope specialist of reasonable expertise. Yet such experts were not consulted, or they were not listened to. The project was eventually scrapped in 2011, wasting more than £100,000 of public money.

The Department for Work and Pensions (DWP) introduced a pilot based on similarly questionable science in 2009, this time of 'voice risk analysis' software designed to assess whether benefit claimants were lying. There was no peer-reviewed scientific evidence that this system was likely to be effective, and the technology had been criticized in the scientific literature as 'at the astrology end of the validity spectrum'. The DWP decided to test it anyway, at a cost of £2.16 million.

The tests failed: of the twenty pilots, only five reported a correlation between claims judged as high risk and actual fraud. As Francisco Lacerda, Professor of Phonetics at Stockholm University, put it: 'I would be surprised if they had reached another conclusion. The problem with this device is that it is not even plausible to begin with. Had the Department asked scientists in the UK they would probably have been advised not to bet on it, so this is a very expensive way of reaching an obvious conclusion.'

Voice risk analysis was at least abandoned before it had any ill effects. But bad science in law enforcement can claim lives, and not only by destroying them as Meadow's false statistics did to Sally Clark.

In 2008, the Iraqi Interior Ministry signed a £19 million contract with a British company called ATSC to supply 800 bomb-detection devices. ATSC claimed that its ADE651 device worked by 'electro-static magnetic ion attraction' and was capable of detecting explosives up to a kilometre away. An array of plastic-covered cards with barcodes is arranged in a wand, without a power supply. The wand is supposed to swivel towards the source of any explosives; the manufacturers also claimed it could detect drugs, truffles and even ivory.

The ADE651 device is essentially a dowsing rod, and, like dowsing rods, it doesn't work. Dale Murray, head of the US National Explosive Engineering Sciences Security Center at Sandia Labs, told the *New York Times* that his group had 'tested several devices in this category, and none have ever performed better than random chance'.

The British government eventually banned the export of the devices in January 2010, after ordering tests that showed 'the technology used in the ADE651 and similar devices is not suitable for bomb detection'. But by then much damage had been done. On 25 October 2009, suicide bombers killed 155 people in Baghdad after getting 2 tonnes of explosive past checkpoints at which the ADE651 was typically used. Had bomb-sniffing dogs been deployed instead, the chances of preventing this attack, and many others like it, would have been incomparably greater.

Dubious technology like the ADE651 should never be allowed to make it into public policy. But want of scepticism is not the only way in which bad science creeps into the law. Politicians who take charge of justice systems also have a disconcerting habit of ignoring scientific evidence that they find inconvenient. And if scientists complain, they like to shoot the messenger.

Politicians who take charge of justice systems have a disconcerting habit of ignoring scientific evidence that they find inconvenient.

The blunt end of justice

It was lunchtime of Friday, 30 October 2009, and Professor David Nutt was preparing his slides for an academic meeting on addiction when his mobile rang. On the line was a journalist from the BBC. The Professor of Neuropsychopharmacology at Imperial College London, who chaired the government's Advisory Council on the Misuse of Drugs (ACMD), had been booked to discuss ketamine on Radio 1 the following Monday, but the Home Office had rung to cancel. The reason, the reporter said, was that Nutt's position was under review.

It was the first Nutt had heard of it. He had met the ACMD secretary and the Home Office chief scientist that morning, who had said nothing to suggest his job was in jeopardy. At 3pm, however, he took another call. Did he have access to his email, the ACMD secretary asked. If so, he ought urgently to read a letter from the Home Secretary, Alan Johnson.

Five minutes later, Nutt logged on to his email. Johnson was asking for his resignation. The scientist told the ACMD secretary he didn't want to resign. 'You've misunderstood,' his colleague replied. 'That's code for "You're sacked."'

The reason for his dismissal was a comment piece Nutt had written for the *Guardian* the previous day, which abridged the Eve Saville Lecture he had given to the Centre for Crime and Justice Studies at King's College London in July. It examined the 2008 decision by Johnson's predecessor, Jacqui Smith, to upgrade the classification of cannabis from class C to class B, raising the maximum penalty for possession from two to five years to 'send a message' about its dangers. Nutt said her action, in defiance of the ACMD's recommendations, had stepped beyond the science and might prove counterproductive. Far from deterring drug use, tougher controls without evidence of harm could undermine trust in the whole system.

'I think we have to accept young people like to experiment, and what we should be doing is to protect them from harm at this stage of their lives,' Nutt wrote. 'We have to tell them the truth, so that they use us as their preferred source of information. If you think that scaring kids will stop them using, you're probably wrong.'

Nutt was reiterating the ACMD's position, which he had spelled out several months previously in the King's lecture. He knew his presentation would be sensitive, so he cleared his slides beforehand with the Home Office chief scientist. The *Guardian* article said nothing he hadn't said before. But to Johnson, Nutt had 'crossed a line' by 'campaigning against government decisions'. He would have to go.

'It is important that the Government's messages on drugs are clear and as an adviser you do nothing to undermine the public understanding of them,' the Home Secretary wrote in Nutt's letter of dismissal. 'I cannot have public confusion between scientific advice and policy.'

Johnson was right to a point: there was confusion between scientific advice and policy. But it was not of Nutt's making. The government's position did not square with the science for a very simple reason: the science had been rejected. Nutt's crime was to refuse to pretend that his committee had never given the Home Office inconvenient recommendations. Unable to erase the ACMD's unhelpful message, Johnson killed the messenger.

His decision betrayed much about the dysfunctional way in which ministers handle scientists and scientific evidence. To Johnson, Nutt was just a subordinate causing trouble, who could easily be removed. That Nutt was supposed to be independent of the government, and that his job was to be true to the evidence, not the policy, did not appear to cross his mind.

The responsibility for deciding policy, on drugs as on other issues, rests ultimately with elected politicians, who are entitled to reject the expert advice they receive. Yet scientists who give their time to guide ministers, often, as in Nutt's case, without payment, deserve to be treated with a modicum of respect in return. They should be consulted in good faith before decisions are made. Their independence and their right to pursue their normal academic work must be guaranteed. And they shouldn't be punished for reaching politically unpalatable conclusions, or for sharing these with the public.

In the Nutt affair, the government failed on all counts.

Inside the Nutt case

As we saw in Chapter 3, the ACMD was asked to review cannabis *after* the Prime Minister had made his determination to toughen its classification clear. The government was obliged by the 1971 Misuse of Drugs Act to seek the council's views, but it had no intention of listening to them if they proved unwelcome. Ministers behaved the same way when they asked the ACMD to review ecstasy in 2008. As the council was still deliberating, Jacqui Smith's spokesperson told the BBC: 'The Government firmly believes that ecstasy should remain a class A drug.' When the ACMD recommended downgrading the drug to class B, it was overruled. Its advice never stood a chance of being considered.

The Home Secretary's second mistake was to fail to recognize that Nutt was not a minister bound by collective responsibility, or a civil servant forbidden from unauthorized public comment about policy. Rather, he had been appointed to chair the ACMD because he was a leading scientist, and that meant he had academic responsibilities to fulfil. In Nutt's case, these included his research into addiction and drugs harm, and publishing and lecturing about his findings. He couldn't be expected to censor himself so his every public pronouncement followed government policy. Yet that is what successive home secretaries expected him to do.

In January 2009, Nutt published a paper in the *Journal of Psychopharmacology* that compared the risks of two popular activities. One was recreational use of ecstasy. The other was what he called 'equasy' – Equine Addiction Syndrome, or a love of horse-riding. The latter, he pointed out, was responsible for at least ten deaths and 100 road accidents each year. As about thirty deaths a year involve use of ecstasy, the risks involved are broadly comparable. Yet only one is controlled as a class A drug.

Jacqui Smith was outraged. 'I had the Home Secretary ringing me up and remonstrating with me about the reprehensible thing I'd done,' Nutt recalls. 'She said I'd let down the families of people taking ecstasy. She kept on telling me you cannot compare illegal actions with legal ones. But why not? Don't we have to make comparisons like that to evaluate whether the harms are sufficient to make something illegal? I was completely stunned.'

Smith clearly expected her adviser to refrain from publishing anything in his field that might conceivably depart from government policy. She showed a complete lack of understanding of how science might generate more informed debate about the risks of different activities. Nutt was being asked to suspend his academic independence and integrity.

In sacking Nutt, Johnson took this insult to another level. Smith's replacement as home secretary was no more obliged than she was to accept the recommendations of the ACMD. But by dismissing him for 'campaigning' when he persisted in communicating the panel's views, he acted as if its members and chairman should have abrogated them.

The affair – inevitably dubbed the 'Nutt case' or sometimes, more profanely, 'Nutt sack' – was a clear demonstration of the way ministers want independent scientific advice if it tells them what they want to hear. It cut to the heart of the tense relationship between science and politics. In their handling of Nutt and the ACMD, Johnson and Smith worked through a checklist of the crimes against evidence-based policy we explored in Chapter 3.

They prejudged the evidence, taking decisions about classification of cannabis and ecstasy before the ACMD's reports were in. They then shopped around for alternative evidence to justify those decisions. And they tried to fix the committee, sacking its outspoken chairman.

Revenge of the nerds

Alan Johnson thought that sacking Nutt would be the end of the matter. So routine did he consider the dismissal that he consulted neither Sir John Beddington, the government's chief scientist, nor Lord Drayson, the Science Minister. But as either might have told him, geeks are growing in political confidence and cavalier ministerial treatment of scientific advice is no longer something they are willing to accept. Five years ago the Home Secretary might have got away with it, but not this time. His actions triggered a furore on a scale that would once have been unthinkable.

Two members of the ACMD, Professor Les King and Professor Marian Walker, immediately resigned in sympathy with their sacked chairman, and another five followed them out in the succeeding weeks. Bloggers rallied to Nutt's support. Scientists who would once have chuntered to colleagues in private lined up to condemn the Home Secretary on television and in the press, aided and abetted by the Science Media Centre.

'It is critical that Chairs and members of independent scientific committees are not just independent of Government but are positively encouraged to provide the best possible interpretation of the available scientific data, whether or not that interpretation fits with the current political view,' said Professor Chris Higgins, a veteran scientific adviser who chaired the government's expert body on mad cow disease. 'They have a duty to communicate their findings, especially when ministers would prefer that they did not.'

Within government, Lord Drayson's consternation at the sacking was revealed when emails he sent to the Prime Minister's office were leaked. Sir John Beddington agreed publicly with Nutt's opinion that alcohol and tobacco are more dangerous than cannabis.

The volume of protest focused public and media attention on the government's misuse of evidence. Drayson and Beddington began to share their fears that experts might begin to refuse invitations to sit on advisory panels unless bridges were mended. With the issue of evidence and scientific advice firmly on the political agenda, here was a chance to draw up some rules of engagement that might help science to find its rightful place in Whitehall. Science's emerging political wing duly swung into action.

A week after the dismissal, Sense About Science organized a statement of principles for the treatment of expert advice, endorsed by dozens of senior scientists and science advisers, which it challenged the government to adopt. It accepted that ministers should be free to reject such advice, but also urged them to guarantee the academic freedom of advisers, the independence of advisory committees and timely ministerial consideration of their recommendations.

The government made a show of rising to the challenge. It accepted most of the Sense About Science proposals, including, importantly, the right of committees and their chairs to communicate their findings to the public even when they are rejected. The principles, now written into the ministerial code, illustrate what geek lobbying at the right moment of government weakness and sensitivity can achieve. But they are undermined by a small phrase that ministers added, insisting that 'Government and its scientific advisers should not act to undermine mutual trust'.

This inchoate provision offers a get-out-of-jail-free card to ministers of a Johnsonian bent. Any adviser who speaks out of turn can be accused of undermining trust and dealt with accordingly. As with the Science is Vital campaign, the geeks' victory remains frustratingly incomplete.

Old habits of evidence abuse, too, die hard in Whitehall, as another episode touched off by the Nutt case attests.

The mass resignations from the ACMD that followed the affair highlighted a flaw in its constitution. The law required the council to include representatives from six specific scientific disciplines, including veterinary medicine and dentistry. When some of these members quit, the ACMD was left inquorate and paralysed. Both scientists and ministers agreed that this was unhelpful and needed reform.

In December 2010, the new Conservative–Liberal Democrat government published a Police Reform and Social Responsibility Bill with a clause that sought to address this anomaly. It didn't do so in quite the way that geeks had anticipated. The legal clause that required the appointment of people with 'wide and recent experience' of the six scientific disciplines was not altered to allow a little more flexibility. It was excised completely, leaving no requirement for any scientists to be appointed at all. Ministers were getting rid of the problem of inconvenient scientific advice by abolishing their obligation to seek it, let alone to listen.

The following month, the Home Office announced new appointments to the ACMD to replace the members who had resigned in support of David Nutt. The choices did not exactly reassure those who were concerned about the government's intentions. One name was Sarah Graham, an acupuncturist. Another was Hans-Christian Raabe, a Manchester GP who belonged to a radical Christian pressure group.

Raabe had previously distinguished himself as the author of a report to the Canadian Parliament that stated: 'Any attempts to legalise gay marriage should be aware of the link between homosexuality and paedophilia.' The government was entrusting drugs policy advice to someone who was quite prepared to quote non-existent evidence to support a religious crusade.

Raabe's appointment, however, was not without its benefits. As soon as it became public, geeks were on the case. Evan Harris, the former MP and champion of science, tipped off *The Times* and the BBC, both of which highlighted Raabe's opinions. Skeptical bloggers such as Tom Chivers and Andy Lewis pored over Raabe's published views and tore into the dubious evidence with which he supports them. 'When David Nutt was sacked for expressing views on relative harm, we were angry that Government was ignoring the advice of its advisors,' wrote Lewis. 'Now we find ourselves hoping they do.'

> *'When David Nutt was sacked for expressing views on relative harm, we were angry that Government was ignoring the advice of its advisors. Now we find ourselves hoping they do' –*
> Andy Lewis

The Home Office soon realized it had blundered. Just two weeks after his appointment, Raabe was sacked for failing to declare his authorship of the paedophilia paper, which 'raises concerns over his credibility to provide balanced advice on drug misuse issues'.

Another small victory – even if the bigger issue of scientific representation on the ACMD remains unresolved.

Making a hash of it

The deaf ear that politicians turn to science is especially unfortunate for drugs policy because science has so much to contribute. The principal justification for infringing individual liberty to control access to certain substances is to limit the harm that they cause, both to the people who take them and to society at large. The level of harm that different drugs inflict, and the best approach to minimizing this, is difficult to determine and there is plenty of room for honest disagreement. But it becomes close to impossible if the findings and methods of science are not given the weight they deserve.

The scientific approach, oddly, has largely been accepted by many governments for a drug that both David Nutt and Alan Johnson would agree is more damaging than cannabis or ecstasy: heroin. There is little dispute about the category in which it belongs: it is the epitome of a class A drug. Yet the state's approach to dealing with heroin addicts is now founded on reducing the harm it causes to users, in line with the best scientific research.

This evidence-based strategy has two key elements. To stabilize heroin addiction, and to reduce the incentives to use crime to fund it, users are prescribed methadone or, increasingly, heroin itself. Needle-exchange programmes are also made available, so that addicts have no reason to share needles and run the risk of infection with HIV or hepatitis.

Scientific evidence has proved the value of harm reduction. Randomized trials show that prescribing heroin to long-term addicts significantly improves both health and social outcomes. Many addicts on such programmes are able to hold down jobs. Criminality is also much reduced, as users no longer need to fund an expensive habit. This makes such programmes cost effective: their expense is more than compensated for by lower levels of crime.

There is good evidence that exchange programmes significantly reduce sharing of needles, and some evidence that this improves

HIV infection rates. Both aspects of harm reduction, moreover, pass a further test that is critical to political acceptability: they do not encourage people to begin using heroin, but rather allow existing users to manage their addiction more safely.

These compelling arguments for harm reduction have not been accepted everywhere. The US retains a ban on federal funding for needle-exchange, even though the National Academy of Sciences recommended as long ago as 1995 that 'well-implemented needle exchange programmes can be effective in preventing the spread of HIV and do not increase the use of illegal drugs', and the Department of Health and Human Services reached a similar conclusion three years later.

Alan Leshner, who was director of the US National Institute on Drug Abuse when the Clinton Administration considered and rejected federal funding of needle-exchange, describes a sequence of events that will be familiar to recent members of Britain's ACMD. 'There was a ban on using federal money for needle-exchange programmes unless we could show that they reduce the spread of infectious disease and do not increase drug use,' he explains. 'We could and did show that: we had thirteen or fourteen studies. The politicians didn't want to listen.'

Barry McCaffrey, President Clinton's 'Drugs Czar', 'went after the studies and tried to present them as bad science', Leshner recalls. The result was that Clinton ducked the issue, allowing 'local communities' to decide whether needle-exchange schemes should be implemented. As Chris Mooney recounts in *The Republican War on Science*, the Bush Administration went further, referring journalists to questionable evidence to support its opposition to harm reduction, and quoting supposedly supportive academics who in fact rejected the President's stance.

In most European countries, though, harm reduction for heroin has become government policy. It makes you wonder why, if a scientific approach is deemed viable for the hardest and most politically sensitive of drugs, similar evidence cannot be brought to bear for less-damaging substances such as cannabis and ecstasy.

It is not as if scientists have not tried. There is evidence there to allow politicians to control these drugs according to harm if they wished to do so.

It is known that cannabis has low toxicity: the statement by Gordon Brown, as Prime Minister, that some powerful strains of skunk are 'lethal' is factually incorrect. The risk of addiction is also low. There is evidence that it can raise the risk of psychotic illnesses such as schizophrenia among some users, but the dangers to individuals are small. Research led by Matthew Hickman, of the University of Bristol, shows that it would be necessary for at least 2,800 cannabis users, and perhaps as many as 10,870, to stop smoking the drug to prevent a single case of psychosis. There is also a paradox that Nutt mentioned in the article that got him sacked: while cannabis use has been increasing in recent decades, the incidence of schizophrenia has been declining.

When compared to legal drugs such as alcohol and tobacco, there is little doubt that cannabis is less harmful. Almost exactly a year after his sacking, David Nutt published a paper in the *Lancet* that assessed the harms caused by twenty drugs according to sixteen criteria, such as addictive potential, known health risks and social impact. Heroin, crack cocaine and crystal meth emerged as the most harmful to individuals; alcohol, heroin and crack cocaine were the most harmful to others. Cannabis scored 20 points out of a possible 100 on both measures; alcohol, the worst drug overall, scored 72.

Nutt accepts that the strength of cannabis has increased with the emergence of potent strains of skunk, but he contends that users have changed their behaviour accordingly. 'It's true that it's different, but that's like saying wine is different from beer,' Nutt says. 'If you drink wine by the pint you'll get more intoxicated, but most people know the difference. Just because skunk is stronger is no reason to suppose it's more harmful. A novice might get more stoned, but users adapt.'

There is no solid evidence as to whether the legal status of the drug effectively deters its use. But Britain's 2004 decision to downgrade its classification, reversed in 2008, did create something of a natural experiment – even though an opportunity to evaluate its impact systematically was missed. The British Crime Survey showed no change in patterns of overall cannabis use, and a trend for slightly lower use among young people that began in 1998 continued. The police estimated that the new regime saved 199,000 hours of officers' time in its first year. At worst, the change was neutral; at best, it saved money, decriminalized otherwise law-abiding citizens and contributed to a downward trend in cannabis abuse.

The sum of this evidence, which convinced the ACMD that cannabis ought to be controlled in class C, has ultimately had less appeal to politicians than other hypotheses with considerably shakier foundations. One of these was the 'gateway theory' of drug abuse. This is the common-sense position that experimenting with 'soft' drugs such as cannabis must be prevented at all costs, even if it does little direct harm, because it serves as a gateway to the abuse of harder drugs. As most heroin users have taken cannabis in the past, cannabis is said to be a risk factor for heroin abuse. Yet there is hardly any evidence in its support.

John Strang, director of the National Addiction Centre, gave an elegant summary of the gateway theory to the Commons Science and Technology Committee when it prepared its 2006 report, *Drugs Policy: Making a Hash of It*. It's true that most heroin users have previously used cannabis, but it has been impossible to show that this correlation implies causation. 'Going to primary school is a gateway to being a heroin addict but you are not implying there is a causal relationship between the one and the other,' he said.

When the ACMD considered the issue in 2002, it concluded: 'Even if the gateway theory is correct, it cannot be a very wide gate as the majority of cannabis users never move on to Class A drugs.' RAND Europe, which evaluated the evidence for the Commons committee, found that 'the gateway theory has little evidence to support it despite copious research'.

Then there is sending a message – another cornerstone of Brown's cannabis strategy. Politicians claim all the time that their decisions on drug classification have an important influence on the social acceptability of drug use, and that tougher restrictions have a deterrent effect. Yet this has barely been researched. 'There is no evidence that classification makes any difference at all,' says Nutt. 'It has never been studied. It's an act of faith. This should be a fundamental question: do these laws do what they're supposed to do? But we don't ask it.'

Fair tests

The effectiveness of drug deterrence is not the only question in criminal justice that might become more solvable if politicians were to bother to use the methods of science to study it. We saw in the last chapter how ministers are fond of following ideology and intuition when embarking on new educational initiatives, but much less keen on evaluating them properly to find out whether they really work. Their habits are no different where crime and punishment are concerned.

In 1998, the Labour government introduced the Drug Treatment and Testing Order (DTTO), an alternative community sentence. If a young drugs offender consents to a DTTO, he must receive compulsory treatment, backed up with testing, for a period of between six months and three years.

The hypothesis that treatment might be more likely to prevent recidivism and improve individual health than prison is perfectly plausible, and the Home Office took the commendable decision not just to introduce the orders, but to run pilot projects to assess their merit first. These pilots were deemed a success, and DTTOs were rolled out nationwide in 2000. Yet nobody knew at that point whether the orders actually helped young offenders to quit drugs and crime, and nobody is any the wiser today. That is because the research intended to evaluate DTTOs was designed in such a way as to be virtually worthless.

When the pilot programme was started, Sheila Bird, of the MRC Biostatistics Unit, immediately sounded alarm bells. The pilots, she pointed out, would include too few young offenders to achieve statistical significance: even if DTTOs were moderately successful and reduced criminality, this positive effect might not be detectable.

There was also no chance of the pilots establishing whether the orders reduced drug-related deaths, a key public health outcome. And perhaps most importantly of all, the research wasn't randomized.

Randomization is one of the most powerful tools science has developed for conducting trials of human subjects, whether those trials are trying to assess the safety or effectiveness of a medicine, a teaching strategy or a new criminal sentence. They add an extra degree of rigour to research. If participants in a study are allocated at random to both the intervention group and the control group, there is minimal room for bias in the selection of either. Without randomization, there always remains a possibility that any differences observed between subjects and controls may be the result of underlying differences between the two groups, rather than an effect of the intervention.

Medical researchers like Bird thus go to great lengths to place randomization at the heart of the trials they design. 'Randomization takes into account not only the influencing factors you know about, but those you don't know about – the unknown unknowns,' she explains. 'At the point of eligibility, you don't know which way the coin is going to fall.'

It would have been a simple matter to randomize the DTTO pilot. When a qualifying offender was convicted, the judge would have assigned him to prison, or to a treatment and testing order, according to a randomly generated code. But the Home Office baulked: sentencing at random was too controversial to contemplate, even with offenders' consent. Decisions about which offenders would receive the experimental orders were instead left to individual judges, who were not even asked to record the sentence they would otherwise have passed. This, Bird points out, compromised the pilots before they had begun. Judges could easily be tempted, consciously or subconsciously, to cherry-pick the less serious offenders, who had a lower risk of recidivism, for one arm of the trial or the other.

The pilots recorded a fall in self-reported drug use and crime among offenders given DTTOs, and they were declared a success. Their lack of statistical power and randomization, however, undermined their ability to reach reliable conclusions. No pharmaceutical company would have got away with running so shoddy a trial. Yet such evidence was sufficient to change sentencing policy to the extent that more than 21,000 DTTOs were handed out to offenders in 2003.

The Home Office learned little from this fiasco. In 2002, it asked Bird, with Gillian Raab and Helen Storkey, of Napier University in Edinburgh, to advise on the design of a pilot of Project Blueprint, a £6 million drugs-education strategy for eleven- and twelve-year-olds. The statisticians advised that at least fifty-seven schools would be needed in both the intervention and control groups to show an effect on any type of drug use. As the most harmful drugs – heroin, cocaine and crack – are used infrequently by teenagers, more than 1,000 schools would be needed in each arm to demonstrate an impact.

These recommendations were ignored. When the pilot went ahead, in 2004 and 2005, the Blueprint scheme was introduced in twenty-three schools, with six others supposedly serving as controls. The evaluation document was published in 2009, and guess what? The samples, it concluded, were too small to generate significant data.

Though the study's shortcomings had been known at the outset, the Home Office proceeded despite knowing it could determine nothing. As Mark Easton, the BBC's Home Affairs Editor, wrote when he revealed the results (which were slipped out without a press release to alert journalists): 'At some point during the years of research, Home Office ministers must have been told of the problem at the heart of the Blueprint project. It would appear they were asked for more money to make the findings robust, but refused. Rather than pulling the plug on the whole evaluation, however, the process was allowed to struggle on in the hope that some broader comparisons might still be valid. It was to prove a vain hope.'

The Home Office is a serial abuser of statistics. In 2008, Jacqui Smith announced a 27 per cent fall in the numbers of people admitted to hospital with stab wounds in nine areas that had been targeted by a knife-crime action plan. She was sternly ticked off by

Sir Michael Scholar, head of the government's statistics watchdog: the figure was provisional and was not supposed to be published as it had not been properly checked.

Easton established that the true picture was worse even than Scholar realized. Department of Health statisticians had informed the Home Office that stabbings had been falling in the target areas before the action plan was introduced. Smith and her officials, however, chose to ignore this, cherry-picking a dubious statistic that appeared to show her policy to have been successful.

Then there is the Social Impact Bond, a flagship project of David Cameron's 'Big Society'. Under this initiative, which is being piloted at Peterborough Prison, a company called Social Finance is putting its own money into a mentoring programme designed to reduce re-offending when prisoners serving sentences of less than a year are released. Its investment, plus a small profit, will be paid back only if re-offending rates actually drop.

It is an admirable idea, described by Michael Green and Matthew Bishop, the authors of *The Road From Ruin: A New Capitalism for a Big Society*, as 'the best Big Society proposal'. The pilot also has some commendable features: success will be measured according to re-offending rates among all released prisoners, not just those who participate in the optional programme, to prevent selection of easy cases.

Yet as Bird points out, it also has an intrinsic weakness. It isn't randomized, which limits its capacity to prove very much.

Peterborough could have been selected at random for the intervention, from a group of similar prisons that will now serve as controls. It wasn't: it was specifically chosen as the study site, and the characteristics that caused it to be chosen could plausibly account for the programme's success or failure.

Neither have inmates been randomly allocated to receive specialist support or to join a control group. Each Peterborough prisoner will instead be 'matched' to up to ten control prisoners serving similar sentences elsewhere – a task that is fraught with potential for bias. A further issue is that Peterborough will benefit from substantial extra resources, which will not be provided to any of the prisons where the control inmates are held. It could be that it's the money, not the mentoring scheme *per se*, that accounts for any benefits that are seen. Other prisons might have done just as well with similarly enhanced budgets spent a different way.

Any data that the study captures risk being too unreliable to show whether the Peterborough programme has been effective, Bird says. At best, the pilot can work only as a proof of concept, showing whether it's possible to deliver an intervention like this. It might well be a brilliant success; it might achieve little. But we aren't going to know either way until we randomize.

'All these problems could have been averted when thinking through how to design these pilots,' says Bird. 'Too many civil servants don't have any experience of commissioning research that's properly designed, and they don't consult people who do. People who design trials for a living know where many of the pitfalls are, and how they can best be avoided. Yet our skills aren't thought to be relevant to criminal justice. That's a shame. If you mess up medical research, you can ruin lives, but the same is true of the justice system.'

Crime scene investigation

There is another way. As a maxillofacial surgeon, Professor Jonathan Shepherd's clinical practice often involves reconstructive work following severe facial injuries. While working in Cardiff in the late 1980s, he became increasingly upset and frustrated by the number of young people admitted to his Accident and Emergency department who had suffered severe cuts in pub brawls. It was an experience that was to inform two pieces of research that provide a model of how science can and should contribute to better crime policy.

Shepherd's first stroke of genius was to recognize that some good could come out of the very hospital admissions he found so disturbing. The pattern of facial injuries he was treating, he realized, represented valuable data that might be of considerable use to the police. Even when potentially disfiguring injuries are involved, people who are glassed in pubs do not always report the incident as a crime. Some know they were not the passive victims of random violence but active participants, perhaps even the instigators of the fights in which they were hurt. Others fear reprisals. Either way, victims can be reluctant to cooperate with the police. But they invariably end up in A&E.

'By matching up police and A&E data, it became clear to me that the police only learn of about 25 per cent of the violence that puts people in hospital for treatment,' Shepherd says. 'To think that there were a whole lot of patients on my operating table whose assailant would never even be investigated let alone be brought to book was a real motivator. The hypothesis I generated was that the use of data from A&E departments might enhance violence prevention over and above the prevention that was possible using police intelligence alone.'

Over almost three years, Shepherd and his colleagues carefully collected details of every case in Cardiff when a victim of violence arrived at A&E. The researchers recorded the precise location of the incident, the day of the week, the time of day and the type of weapon involved. This information was then shared with the police and the city council, allowing them to draw up a detailed map of violence hotspots. A disproportionate number of incidents turned out to be occurring in a few particularly unruly establishments, which the police were then able to target for especially close attention. In some cases, Shepherd's data were instrumental in the withdrawal of licences from problem pubs and clubs.

Shepherd's team wasn't content just to assume that their strategy was working, however. They also determined to evaluate it properly, to put it to the test.

A randomized trial wasn't possible on this occasion: Shepherd's practice and police contacts were in Cardiff, so that city had to be chosen as the intervention group. So the researchers did the next best thing, using demographic data to identify the most similar British cities and then using them for comparison.

The results, published in the *British Medical Journal* in June 2011, were astonishing. Over the fifty-one months during which the data-sharing strategy was analysed, hospital admissions due to violence in Cardiff fell by 42 per cent compared to the control cities. In Cardiff, such admissions fell from seven a month to five a month per 100,000 inhabitants. In the comparison cities, the number of such incidents per 100,000 people increased from five to eight per month. The number of minor assaults recorded as crimes also increased in Cardiff while falling elsewhere, suggesting that data-sharing was helping the police to keep track of a higher proportion of violent incidents.

Another insight that emerged from Shepherd's experience as a facial surgeon was how much damage was caused by beer glasses that smash to create a sharp edge. The answer, he reasoned, was the introduction of toughened glasses that smash into tiny and harmless pieces like those of a shatterproof car windscreen. Once again, he didn't simply back his hunch, but instead designed proper experiments.

First, his team asked glass manufacturers to supply him with batches of their wares, which he tested in the laboratory. One type, which had been tempered in the manufacturing process, turned out to be six times harder to break than the others. Then Shepherd designed an RCT.

'We replaced the entire pint pot stock of fifty-seven licensed premises in the West Midlands and South Wales with tempered or non-tempered glass,' Shepherd recalls. 'The premises were selected at random. Then 1,200 bar staff gave us regular reports on breakages and injuries. The tempered glasses were associated with a lower incident risk of about 60 per cent, including accidental breakages and assaults. It meant that we got an evidence-based intervention. I campaigned for a wholesale switch, and after it finally happened, in 1998, the British Crime Survey demonstrated a significant reduction in the number of glass-related assaults.'

The dispassionate eye of the RCT smoked out an effect with great relevance to injury prevention that would otherwise have remained hidden. The work won Shepherd the 2008 Stockholm Prize for Criminology, often regarded as the Nobel of the field.

'This isn't high-level strategy, it isn't particularly hard,' Shepherd says. 'It's about taking your experience from the coalface of public service, and coming up with ideas that you put to the test. This sort of R&D ought to be deeply embedded into public service everywhere, not least in criminology where it has such potential to do good. I find it extraordinary that so many public services have nothing of the sort.'

Shepherd's track record of using science to make violence prevention somewhat more tractable has won him the confidence of the South Wales Police, who regularly seek his advice. The force has also established a criminology school, in collaboration with the University of Wales, that develops better ways of introducing scientific methods into policing. His experience shows what can be done when people who understand and value what evidence can bring to public policy share their skills with the appropriate public services.

Unfortunately, scientists who seek to create such partnerships with the architects of public policy will often find themselves rebuffed. We will hear, as Sheila Bird did, that our methods are inappropriate, impractical or even unethical. Not everyone who knows something of the value of RCTs is as well placed as Jonathan Shepherd to deploy that knowledge to such great effect, and it took years before Shepherd was taken seriously by the police.

What geeks can always do, however, is to critique the evidence that is used to justify criminal justice policy, as Bird has done for so many of the Home Office's botched pilots. Blog about their failings. Go to the media: even journalists who might seem less than sympathetic understand that government failure always makes a good story. If we can embarrass ministers and civil servants often enough about their refusal to use the best tools available to evaluate policy initiatives, perhaps they will start to realize that generating better evidence is neither impossible nor unhelpful.

Above all, these are issues that need to be placed on, and kept on, every politician's agenda. Many MPs are actually quite receptive to scientific arguments about drugs control – at least when they are not in office. Chris Mullin, the former Labour MP, recalls in his diaries the contribution of a young Conservative backbencher to a Home Affairs Select Committee inquiry that recommended a harm-reduction approach to cannabis control. His name was David Cameron, and he has become strangely silent on the subject of late.

Politicians like Cameron revert to old-fashioned 'tough on drugs' language at the first sniff of office because they know there are substantial costs to being open-minded about the evidence. Flirt with science-based policies, and your political opponents and much of the media will paint you as soft on drugs and crime. So they play it safe, ruling out approaches that might actually stand a chance of doing some good.

That's not going to change until those of us who understand what following the evidence might contribute make as much fuss as the *Daily Mail*. Geeks need to show politicians and their parties that it isn't only sound policy that suffers when they reject science. It can be bad politics as well.

That's starting to happen. The strength of the backlash against David Nutt's dismissal and the mismanagement of the ACMD took politicians of all parties by surprise. It created a political cost where once there was none, which ministers know will be there again if they persist in business as usual. Our challenge now is to make that cost so steep that politicians no longer wish to pay it.

In each of the last two chapters, we've met ways in which politics and the public services can learn much and more from medicine. But that doesn't mean it can be spared the geek eye. Doctors are better than most other professions at putting scientific thinking into practice. What about the polititions they answer to?

PLACEBO POLITICS

Why science matters in healthcare

IT WAS AN EMINENTLY SENSIBLE PUBLIC HEALTH INITIATIVE — CLEAR, succinct and impeccably founded in the available evidence. 'Alternative remedies which could be dangerous are being targeted by the Government in a major drive to improve health and welfare,' read the departmental press release.

John FitzGerald, the civil servant responsible for the campaign, explained why the British authorities were taking the misleading claims made for many alternative medicines, particularly homeopathy, so seriously. 'Some of these products are claiming to be effective and safe when no scientific evidence has been presented to us to show they are,' he said. '[Citizens] have a right to know if a product does what it claims. The products claim to treat diseases which can cause serious welfare problems and in some circumstances kill . . . if not properly treated.'

To lay claim to a medicinal indication without evidence was at best a deception, a way of making money under false pretences, the government had decided. At worst, it could be dangerous: while homeopathic pills might be harmless in and of themselves, they could encourage the use of ineffective treatments for serious conditions in place of proven ones. If manufacturers of homeopathic remedies were unable to present data showing the safety and effectiveness of their wares, they would have to label them accordingly as devoid of medical benefit. Homeopathy wouldn't be banned, but ordinary people, the government was clear, had a right to a properly informed choice.

This was just the sort of public-health advice for which geeks have been calling for years. Homeopathy is a system of medicine founded on faith rather than science, and while few of its critics wish to see it outlawed, those of us who care about the scientific method think it should be advertised to the public as what it is, not what its proponents would like it to be. But there was a catch. The ministry that issued this press release was the Department for Environment, Food and Rural Affairs (Defra), and FitzGerald is director of operations at the Veterinary Medicines Directorate. The word 'citizens' appears in square brackets in the quote above because the word that was actually used was 'animal owners'.

This admirable clampdown on quack cures was designed to protect pets, not people.

It is rather a different story for human medicine. While the evidence base for homeopathy is so dubious that senior vets now consider its use unethical in animal medicine, another branch of government has deemed it perfectly acceptable for health practitioners to use identical techniques for treating patients. Not only does the Department of Health fail to insist on the evidence-based labelling requirements that Defra demands for veterinary medicine; the agency that regulates pharmaceuticals has explicitly exempted homeopathy from tough standards that apply to conventional drugs. Worse, the state implicitly endorses its use by paying for thousands of patients each year to be treated with homeopathy on the NHS.

We usually expect humans to enjoy better standards of medical care than animals. Drugs, indeed, must be tested on animals before they can be given to people. For homeopathy, the situation is

While the evidence base for homeopathy is so dubious that senior vets consider its use unethical in animal medicine, government has deemed it perfectly acceptable for health practitioners to use identical techniques for treating patients

reversed. Animals and their owners are better protected against quackery than are patients and consumers.

In the previous two chapters, we saw how the machinery of government fails to make use of the methods of science to explore the effectiveness of different approaches to teaching children and controlling crime. In healthcare the problem is a little different.

While the medical profession's record of evidence-based practice is not without blemish, doctors are streets ahead of their colleagues in law enforcement and education when it comes to evaluating the tools of their trade in a rigorous fashion. Drugs are tested first on animals and then in randomized controlled trials (RCTs); regulators demand such evidence before new medicines are awarded a licence. These trials, in turn, are collated and appraised in systematic reviews, which summarize the current state of scientific knowledge. As a result, medical science develops a better understanding of which medical practices work, which do not, and which remain uncertain.

Yet as homeopathy and other aspects of alternative medicine demonstrate, that understanding does not always inform public policy as it should. It is not that these therapies have not been properly researched, but that the results of that research are ignored by ministers and civil servants.

No harm and no good

Homeopathy is a system of medicine first devised in the 1790s by a German doctor named Samuel Hahnemann. Cinchona bark, which contains quinine, was known even then to have some beneficial effects against malaria, and when Hahnemann took it himself, he developed a fever and palpitations. Those symptoms, he thought, were similar to those of malaria and from this single experience he proposed his 'Law of Similars'. Like cures like, he suggested, so medicine should proceed by treating disorders with substances that produce similar symptoms.

To that, Hahnemann added a second hypothesis: that these agents must be diluted for medical use. Under this 'Law of Infinitesimals', one part of the supposedly active ingredient must be mixed with 100 parts of water, then banged hard against an elastic surface in a process known as succussion. These dilutions and succussions are repeated again and again, with every dilution held to increase the potency of the remedy. The dilutions are measured according to the centesimal, or C, scale: for a 2C preparation, the one-in-100 solution is itself diluted with 100 parts of water, so that there is one part active ingredient to 10,000 parts water. At 12C, the remedy becomes so dilute that it is unlikely to contain even a single molecule of the original active ingredient: it is just water.

Hahnemann favoured 30C dilutions for most purposes, leaving one part of active ingredient in 10^{60} parts of water – that's a one followed by 60 noughts. To ingest one molecule of the active ingredient at this dilution, it would be necessary to consume 10^{41} pills – a trillion times the volume of the earth. Oscillococcinum, a homeopathic remedy for flu made from duck liver, is sold at a dilution of 200C. To get a single molecule of that, you'd have to ingest 10^{400} molecules of solvent. There are only about 10^{80} atoms in the observable universe. At such dilutions, it might not surprise you to learn that only a single duck need be slaughtered to keep the world in oscillococcinum for a year.

These core principles of homeopathy have no basis whatsoever in science. The Law of Similars is a misnomer, an example of Richard Feynman's 'cargo-cult science', in which scientific language is co-opted to give credibility to pursuits that eschew scientific rigour. Proper scientific laws, such as the Laws of Thermodynamics, are universal, never broken in the repeatable experiments conducted to date. The Law of Similars is breached time and again by the most successful drugs of modern medicine. Antibiotics are in no sense similar to the bacteria they attack. Platinum-based chemotherapy works because it is toxic to cancer cells, not because it is the same.

Vaccines, it is true, do work by exposing the body to weakened versions of germs, so the immune system can learn to recognize and eliminate them. But the second principle of homeopathy, the Law of Infinitesimals, precludes homeopathic remedies working by this means. To induce immunity against an agent, you have to be exposed to it, and homeopathy exposes the body to nothing but water.

Real scientific laws show that there is a limit to which you can dilute a substance while retaining some of that substance in the resulting solution. The limit is related to a concept called Avogadro's constant, and it corresponds to homeopathic dilutions of 12C. At 'strengths' of 30C or above, homeopathic preparations do not even contain any of the original *water* used in the first dilution, let alone any active ingredients. There's simply nothing in these remedies that could potentially have a medicinal effect.

'Molecules are very small,' explains David Colquhoun, Professor of Pharmacology at University College London. 'A single 300 milligram aspirin tablet contains a billion trillion (10^{21}) aspirin molecules. In contrast, most homeopathic pills contain zero molecules.

'How do we know this? It was all worked out in the nineteenth century. Avogadro's constant is the number of molecules in 1 gram of hydrogen, or in 180 grams of aspirin, or in 342 grams of sugar. (The number of grams is just the mass of a molecule relative to that of a hydrogen atom.) It is a very large number, about 600 thousand million trillion, or 6.023×10^{23}. Although the number is huge it is still tiny compared with the 10^{60} dilution that is commonly used by homeopaths. They call it 30C, which disguises the fact that 10^{60} is a trillion trillion trillion times bigger than Avogadro's number. That means that a 30C pill would have to have a diameter equal to the distance from the Sun to the Earth if it were to contain even a single molecule of the active principal.

'That's how we know that arnica 30C pills contain no arnica whatsoever. Normally such mislabelling would be illegal, but an exception is made for homeopathy.'

Homeopaths like to claim that the water in their remedies somehow 'remembers' the molecules with which it has been mixed to induce a healing effect that science has yet to explain. There is no experimental evidence to suggest that this is so, and much to indicate it is impossible. Clusters of water molecules exist for just trillionths of a second before they are broken apart by Brownian motion. If water can remember, then a fundamental principle of physics that has been confirmed again and again by experiment is wrong. As Colquhoun told the Commons Science and Technology

Committee when it reviewed the evidence for homeopathy in 2010: 'If homeopathy worked, the whole of chemistry and physics would have to be overturned.'

'A 30C [homeopathic] pill would have to have a diameter equal to the distance from the Sun to the Earth if it were to contain even a single molecule of the active principal' – David Colquhoun, Professor of Pharmacology at University College London

Its scientific basis might be preposterous, but in the late eighteenth and early nineteenth centuries homeopaths for a time achieved results that were rather better than those of the conventional medicine of the day. While the bloodletting and purging employed by orthodox doctors were often actively damaging to their patients, Hahnemann's magic remembering water at least fulfilled the first principle of the Hippocratic Oath and did no harm. Homeopathic remedies contributed no pharmacological benefit, but did not compromise patients' chances of a natural recovery, while helping doctors safely to harness the healing power of the placebo effect.

As conventional medicine began to learn from its mistakes and to use the methods of science to evaluate the safety and effectiveness of techniques old and new, this competitive advantage of homeopathy soon evaporated. Once those that were harmful and ineffective were jettisoned, drugs that contain something biochemically active were always likely to out-perform water. So RCTs have consistently shown. While some isolated studies have appeared to show homeopathic pills to be a little better than inert placebos, the overwhelming majority do not.

Systematic reviews, which consider all the RCTs that meet sufficient standards of quality, have found against homeopathy time and again. The Cochrane Collaboration, a clearing house for systematic reviews, has found no evidence to support homeopathy as a treatment for asthma, dementia, attention deficit/hyperactivity disorder or induction of labour, or for palliative care in cancer. Edzard Ernst, Professor of Complementary Medicine at the University of Exeter, has shown that the more rigorous the methodology of the trial, the less likely it is to suggest that homeopathy works.

Homeopathy is not only a system of medicine that lacks a plausible mechanism to explain why it might work. It also fails to produce any results that need explaining. All it does for patients is to invoke placebo effects, which real medicines do just as well as imaginary ones. Homeopathy is not unproven, like many of the teaching techniques and sentencing policies that ministers so stubbornly refuse to evaluate. It is disproved.

Other disproved medical techniques, such as prophylactic removal of wisdom teeth, have been eradicated from state-supported healthcare in line with the evidence. Drugs that fail clinical trials are denied a licence, and approved drugs that turn out to be risky, like Vioxx and Avandia, are withdrawn. Homeopaths, however, get to play by different rules.

A poor prescription

The government knows perfectly well that homeopathy does not work. When the Commons Science and Technology Committee asked Mike O'Brien, the Labour Health Minister, whether he had any evidence that homeopathic remedies were better than placebos, he replied: 'The straight answer is no.' Sir John Beddington, the government's chief scientist, was more explicit. 'I have made it completely clear that there is no scientific basis for homeopathy beyond the placebo effect and that there are serious concerns about its efficacy,' he told the committee. 'There is a danger that the public will think that there is real efficacy for some serious conditions and I believe we have to work on that and make clear that this is not correct.'

For most medicines, this would be more than sufficient to prevent their use by the NHS. The government, however, has decided that a lack of evidence is no reason either to stop doctors from prescribing homeopathy or to stop paying for it with taxpayers' money. 'We should not take the view that patients should not be able to have homeopathic medicine when they want it,' O'Brien told the committee. All parties seem to go along with this: when the Conservative–Liberal Democrat coalition took office in May 2010, the government's position remained unchanged. Anne Milton, the new Tory Health Minister, flatly rejected the committee's recommendation that the NHS stop funding homeopathy. 'The over-riding reason for NHS provision is that homeopathy is available to provide patient choice,' the government said.

Homeopathy doesn't cost the taxpayer a lot of money. While the government does not audit such spending, the British Homeopathic Association estimates NHS funding at about £4 million a year. The four NHS homeopathic hospitals consume further funds – the Royal London Homeopathic Hospital was refurbished recently at a cost of £20 million – and occupy premises that are an opportunity cost. But even this small spend matters because health funding is a zero-sum game. Healthcare has to be rationed somehow, and every pound squandered on a homeopathic placebo is a pound that cannot be spent on treatments of proven effectiveness that are denied to patients on grounds of cost.

Some NHS trusts refuse to provide infertile patients with the IVF treatment recommended by the National Institute for Health and Clinical Excellence, or cancer patients with drugs that extend life-span, while allowing GPs to offer their patients homeopathy. There is a similar opportunity cost in research. In the US, the National Center for Complementary and Alternative Medicine (NCCAM) has an annual budget of about $130 million. This money is ring-fenced for investigating medical hypotheses that would never normally make it past the rigorous peer-review of its parent body, the National Institutes of Health. The result is to divert scarce research funds from studies with genuine medical promise to alternative approaches that are much less likely to yield any benefits.

More important than the size of these bills is the principle at stake. By allowing the NHS to waste any money at all on an alternative medicine that has been disproved, the government sends a message that using it is somehow worthwhile. The NHS is a powerful brand, which gives a stamp of authority to the treatments it adopts. By endorsing homeopathy, it encourages its use and helps the private therapists who sell it to market themselves.

NHS homeopathy also implies that evidence does not really matter. Evidence-based medicine is one of the crowning achievements of twentieth-century science, underpinning fantastic progress in treatment and prevention of trauma and infectious disease, mental illness and cancer. Yet by suspending its standards for homeopathy, the NHS suggests doctors can take evidence or leave it.

Homeopaths get further official support from the Medicines and Healthcare products Regulatory Agency (MHRA), the body that is responsible for licensing drugs. 'Whether it's a medicine you buy, or one prescribed for you as part of a course of treatment, it's reassuring to know that all medicines available in the UK are subject to rigorous scrutiny by the MHRA before they can be used by patients,' the agency's mission statement declares. 'This ensures that medicines meet acceptable standards on safety, quality and efficacy.'

This rigorous scrutiny usually requires the manufacturer of a drug to present a detailed dossier of information from RCTs, to demonstrate that it is safe and effective. Unless, that is, the drug in question is homeopathic. In 2006, the MHRA changed its rules so that suppliers of homeopathic remedies could make medical claims without any supporting scientific data. All they have to do is to show that a product has traditionally been used for a particular purpose and that it has been evaluated by homeopathic 'proving'. Such 'provings' are as misleadingly named as Hahnemann's laws: they do not provide any medical proof at all. They are nothing more than observations, by homeopaths, that certain substances appear to produce the symptoms of diseases and thus might be suitable for treating them under Hahnemann's system. They wouldn't do as evidence in any other medical context.

By accepting evidence from provings, the MHRA allows homeopathic remedies to be sold with profoundly misleading labels. In

May 2009, for example, it approved a label that described 30C homeopathic arnica as: 'A homeopathic medicinal product used within the homeopathic tradition for symptomatic relief of sprains, muscular aches and bruising or swelling after contusions.' Had this been a conventional drug, the manufacturer would have had to present evidence that it actually offers symptomatic relief of sprains, muscular aches and bruising or swelling. To make similar claims for homeopathy, no such evidence was required. Very convenient for the homeopaths, who have no data of the sort.

The MHRA has justified the exemption on the grounds that by approving homeopathic products, it can at least check that they are manufactured according to reasonable quality standards. Michael Baum, Professor Emeritus of Surgery at UCL, put it another way: 'This is like licensing a witches' brew as medicine so long as the bat wings are sterile.'

The Commons Science and Technology Committee agreed with Baum, insisting that homeopathy ought to be held to the same standards as other medicines. The government did not. 'If regulation was applied to homeopathic medicines as understood in the context of conventional pharmaceutical medicines, these products would have to be withdrawn from the market as medicines,' it said. Quite. It is only by breaking the rules that homeopathy can be marketed as medicine at all. If a remedy is labelled as a 'medicinal product' approved by the MHRA for certain symptoms, ordinary people cannot be blamed for assuming that it has actually been shown to be effective against such symptoms. The government is conniving in the deceit of the public.

Diluting the nonsense

Why, then, do politicians pander to homeopaths? A minority are active enthusiasts who will never care much about the evidence, preferring to base their support for it on faith and anecdote. We have already met David Tredinnick, Parliament's most energetic homeopathic pill-pusher, but there are others in higher places with an axe to grind.

Peter Hain, a Cabinet minister under Labour, thinks that homeopathy cured his son's eczema and has promoted it assiduously ever since. As Northern Ireland Secretary in 2007, he launched a £200,000 project to allow GPs in Belfast and Londonderry to refer more patients for homeopathy and acupuncture.

Anne Milton, the Tory Health Minister who responded to the select committee report, believes her experience as a nurse – and her grandmother's experience as a homeopathic nurse – taught her that homeopathy is effective.

In the US, the NCCAM is a pet project of Senator Tom Harkin, who secured funding for it because he thinks his allergies were cured by an alternative therapist who was later forced to stop making false claims for his bee-pollen supplements by the Federal Trade Commission. In 2009, he betrayed his abject misunderstanding of how science works, indeed what it is for, when he professed his disappointment at the Center's results. 'One of the purposes of this center was to investigate and validate alternative approaches,' he told a Senate hearing. 'Quite frankly, I must say publicly that it has fallen short. I think quite frankly that in this center . . . most of its focus has been on disproving things rather than seeking out and approving.' He cared only for science that happened to support his views.

Most politicians, though, are not true believers. Their sympathy for state-sanctioned pseudoscience comes about not because they particularly care for it, but because they have been cleverly lobbied. Alternative therapy has a well-organized political wing, led by no less a figure than the Prince of Wales, and it uses every weapon in the lobbying armoury to press its claims to be a valid branch of medicine that deserves NHS support.

The Prince uses his privileged position to bend the ear of ministers and officials. The MHRA has acknowledged receiving at least seven letters from him about the regulation of alternative medicines. The Prince reportedly buttonholed Andy Burnham, then the Health Secretary, on the subject at a reception in 2009, and recently met Andrew Lansley, Burnham's Conservative successor, to discuss it. This royal lobbying is also opaque: the royal family's communications with ministers and officials are

194 THE GEEK MANIFESTO

exempt from disclosure under the Freedom of Information Act.

Prince Charles's Foundation for Integrated Health campaigned for government registration of alternative practitioners, so they can claim to be state-sanctioned, and for greater use of alternative medicines by GPs. It persuaded the Department of Health to give it £900,000 to produce a patient guide to alternative therapies that contained 'numerous misleading and inaccurate claims concerning the supposed benefits', according to Edzard Ernst. In 2005, it commissioned the Smallwood Report into alternative medicine, which argued that wider NHS provision would save money. Professor Ernst described its conclusions as 'outrageous and deeply flawed'.

The Foundation has been ruthless with its critics. After Professor Ernst criticized the Smallwood Report, Sir Michael Peat, the Prince's private secretary and the Foundation's acting chairman, made an official complaint to Exeter's Vice-Chancellor which nearly resulted in the scientist's dismissal and the closure of his unit. The Prince's Foundation is now defunct, closed in 2010 when its finance director was arrested and subsequently convicted for fraud. Its campaigning now goes on in a new guise: other directors have started another body, which they had the gall to name the 'College of Medicine', to 'take forward the vision of HRH the Prince of Wales'.

The alternative health lobby also understands how politics works, and that many MPs can easily be convinced to take sides by a heavy volume of correspondence from their constituents – especially on issues on which they have no strong convictions. It thus encourages its followers to make their case loudly and often to their elected representatives, many of whom are persuaded to look favourably on what they imagine to be a popular cause with a groundswell of public support. Few MPs have the scientific understanding to question the anecdotal 'it worked for me' stories that advocates of complementary medicine tend to use to promote wider state provision. Even those who are minded to be skeptical often think twice when they realize that many of those who lobby for alternative health are quite prepared to cast their votes on this single issue.

The power of this lobbying is nicely demonstrated by an Early Day Motion proposed to the House of Commons in 2007 by Rudi Vis, then Labour MP for Finchley and Golders Green – Margaret Thatcher's old seat. The motion offered support to the NHS's homeopathic hospitals and extolled complementary medicine's 'potential to offer clinically-effective and cost-effective solutions to common health problems faced by NHS patients, including chronic difficult-to-treat conditions such as musculoskeletal and other chronic pain, eczema, depression, anxiety and insomnia, allergy, chronic fatigue and irritable bowel syndrome'. It expressed concern that some of these hospitals were facing an uncertain future and called on the government to back 'these valuable national assets'. It was signed by 206 MPs, almost a third of the House, including Simon Burns and Paul Burstow, who are now health ministers in the coalition government.

Early Day Motions never become law. Sometimes ridiculed as 'parliamentary graffiti', they are statements of intent to which any MP can put his or her name. The level of support they receive, however, is often revealing of the mood of the Commons. This one was particularly telling, because Tom Whipple, a journalist, asked all the signatories why they had given the motion their support. While fewer than half the MPs replied, the responses of those who did clearly demonstrated the importance of constituency pressure.

A handful of MPs based their support on their own belief in the healing power of homeopathy. More, however, spoke of other reasons. 'My support for homeopathic hospitals is based upon the experience of my constituents as recounted to me,' said Christopher Chope, the Tory MP for Christchurch. Richard Benyon, the Tory MP for Newbury, cited 'constituents who have benefited from such treatments and ... constituents who are practitioners'. Andrew George, the Liberal Democrat MP for St Ives, said: 'In the case of two conditions (eczema and chronic fatigue) constituents are convinced that there was a clear cause and effect.' Most of the signatories had succumbed to the postbag effect.

Politicians fail to take on homeopathy because most of them lack the science to see through it, and because their experience tells them that while there are no votes in insisting on evidence, there are plenty to be lost. Evangelists for homeopathy might not be particularly numerous, but the responses Whipple collected show that they write to their MPs and attend constituency surgeries to share the strength of their views. The alternative health lobby does a better job than its opponents of making its presence felt on the political stage. The result is that the vast majority of MPs who have no strong opinions either way take the path of least resistance, which is to leave NHS provision of homeopathy and the lax MHRA regulations exactly as they are. Why bother antagonizing potential voters when there is no need?

This assiduous lobbying is something with which skeptics are going to have to keep pace if we're to stand a chance of persuading the government to listen to the evidence. Our senior scientists and doctors – the heads of the Royal Society and the medical Royal Colleges – need to use their access to ministers and permanent secretaries as Prince Charles does, to point out the folly of spending scarce public resources on treatments we know to be placebos.

Ordinary geeks must play a part too, apprising our elected representatives of our strong views on the subject. We need every personal anecdote that an MP hears in support of homeopathy to be balanced by a letter or email explaining the ridiculous science behind it and the extensive evidence that it is worthless. We should challenge those who express support for pseudoscience, as Whipple did, to account for their views. Only when politicians know they have constituents who take this issue seriously will more than a handful of them start to take it seriously themselves.

There are signs that this is starting to happen. In 2010, the British Medical Association – the doctors' trade union – voted through policy positions opposing NHS funding of homeopathy and supporting clear labelling of homeopathic remedies as inert placebos. Skeptical-Voter.org keeps a record of MPs' public statements about homeopathy (among other issues); it uses the 'wiki' approach popularized by Wikipedia, so anyone can add new information. And more and more geeks are finding other innovative ways to make evidence count.

Overdosing on water

Boots the Chemist is usually pretty busy on Saturday mornings, but the crowds that gathered outside its branches on 30 January 2010 weren't made up of ordinary customers. These shoppers were geeks. At 10.23am, each of them opened a bottle or bubble-pack of own-brand homeopathic medicine bought from the store and began to scoff the contents. A mass overdose was under way.

Andy Lewis chose homeopathic sleeping pills. 'Right now, if the homeopaths are correct, I should have paralysed arms, be in severe pain, have convulsions, delirium, skin itching all over and be unable to stand,' he wrote in a blogpost set to go live at 10.23. 'That is because I have taken a massive overdose of the homeopathic remedies: Belladonna 30C, Sulphur 30C and Lachesis 5MM.' Michael Marshall, who organized the events, wolfed down a packet of homeopathic arsenic. Dave Gorman, the comedian, took homeopathic arnica, saying, 'I don't know what you're taking, but I will never bruise again. We can test that by punching me afterwards. I volunteer for that in the name of science.'

All lived to tell the tale. The day's only casualty was a nosebleed. 'We're pretty sure it wasn't related,' Marshall says.

Marshall, a marketing manager from Liverpool, devised the protest after admiring the bloggers who took up online cudgels for Simon Singh and exposed the dubious medical claims made by many of the chiropractors who were suing the science writer for libel. If the government wasn't doing anything to expose the dodgy science behind homeopathy, he reasoned, it was up to people like him. At first he investigated leaving information leaflets in GPs' surgeries, but those he contacted turned him down. Instead, he decided to target Boots, whose standards director, Paul Bennett, had recently admitted to a Commons select committee that the high-street chemist sold homeopathic remedies even though he knew of no evidence that they were effective.

First came an open letter, then, when discussing Boots's response in the pub with his friend Mike Hall, the pair struck up a conversation with some visiting Belgians. A mass homeopathic overdose in their country had recently made national news, they told him. Why not stage something similar here, as a simple demonstration of the fact that homeopathy, quite literally, has nothing in it?

Marshall named the campaign 10:23 in honour of Avogadro's constant – 6.023 x 10^{23} – the concept which explains why homeopathic remedies contain not a single molecule of their supposedly active ingredients. Then he began to spread the word. 'The response was staggering,' he says. 'Hundreds of people turned up, and we got an incredible amount of press coverage. Of course, the homeopaths all said it was a stunt, that it didn't prove anything, and they're right. We didn't need to prove anything – that's already been done. What we did was to highlight the issue, to bring it to people's attention.'

Marshall talks proudly of some of the emails and letters he had from ordinary people, telling him they hadn't realized that homeopathy lay outside the medical mainstream until they encountered his campaign. An analysis by David Waldock, of the Open University, found that in the months following 10:23 and the publication of the Commons select committee report, media coverage of homeopathy shifted in emphasis to questioning the ways it is regulated and its provision by the NHS.

Boots hasn't stopped selling its homeopathic remedies. Yet the stunt is just the sort of thing geeks should be doing if we're to change the way homeopathy is perceived by the public and politicians, and indeed by the retailers that happily promote such bogus treatments. It might have been what Mike McRae, author of *The Tribal Scientist*, describes as a 'placebo protest': an event that accomplishes little directly, but which makes participants feel better about themselves. But placebo protests have a wider utility, as masters of the publicity stunt like Greenpeace well understand. They can energize protesters, motivating them to take further action and, by making a case more visible, they make it harder to ignore.

There's a challenge here for the organizers and supporters of 10:23. Fun as crunching sugar pills outside a high-street chemist might be, we can't just leave it at that. People who enjoyed taking part in 10:23 also need to petition and lobby. We must convince politicians and regulators that 'it worked for me' enthusiasts aren't the only people with a position on homeopathy that might affect their vote.

We could also help out another group of geeks who are taking on pushers of pseudoscience with a different strategy. The government, as we've seen, is reluctant to bring in new measures to protect the public from misleadingly sold alternative medicines. But consumer protection legislation that is already in existence provides more room for taking on dubious claims than those who make them usually appreciate. Some skeptical bloggers have begun to use it. The British Chiropractic Association is no longer the only body that is starting to feel the quacklash.

Besides its homeopathic range, Boots stocks and promotes many other supposedly medical products that lack evidence of effectiveness. There is, for example, the 'Ladycare menopause relief magnet'. This device, which you're supposed to pop into your pants, purports to 'reduce or completely eliminate symptoms of menopause'. There's no robust evidence that it works: as with homeopathy, the proposed physiological mechanism is dubious, and the manufacturers quote anecdotal reports by satisfied users, not the results of proper randomized trials. Boots, though, is happy to sell it anyway. It costs £20.45 on the company's website.

To Simon Perry, a computer programmer from Leicester who runs the city's Skeptics in the Pub group, this nonsense amounted to advertising under false pretences. The 'fanny magnet', as he nicknamed it, was being offered as part of a three-for-two promotion on the Boots website, and that meant it fell under the remit of the Advertising Standards Authority (ASA). As the ASA's code of practice requires that advertisers must be able to substantiate any claims made in promotional material, Perry saw an opportunity. He fired off a complaint about the Ladycare device. For good measure, he also reported another 239 products in the online promotion.

Boots might have been able to turn a blind eye to the 10:23 protests, but complaints to the ASA were a different matter. Ofcom, the media regulator, devolves its statutory responsibility for advertising to the ASA, which is obliged to investigate complaints from the public. Its rulings cannot simply be ignored. When the

ASA contacted Boots about Perry's complaint, the chemist agreed to withdraw more than sixty of the products he had highlighted from its promotion and to remove many of the dodgy health claims. In the case of the fanny magnet, Boots eventually went so far as to photoshop the packaging, so that none of its purported benefits any longer appeared on the website. It continues, though, to sell a device that it knows makes claims that cannot be justified.

Perry didn't stop at this partial victory. The ASA has no power over Boots's in-store promotional material, in which it made assertions about the health benefits of homeopathic products, but the MHRA does. Despite the lax rules about homeopathy discussed above, it also runs a different 'simplified scheme' for licensing some homeopathic remedies, which doesn't permit advertising of any medical claims. When Perry found a point-of-sale display that appeared to breach these rules, by recommending remedies licensed under the simplified scheme for allergies, infections, insect bites, headaches and earaches, he complained. In November 2011, the MHRA ruled in his favour and Boots withdrew its misleading advertising.

Perry has shown how determined activism that makes use of existing consumer safeguards can make a difference. He hasn't yet persuaded Boots to stop selling snake oil, but he's given it a cost. The retailer has had to expend time and effort complying with regulations, and it has received negative publicity in the press. Our challenge now is to help people like him to increase that cost to a point at which Boots no longer thinks it worthwhile.

To that end, Perry has written a software package called FishBarrel, which trawls the internet for health claims made by alternative medical practitioners and pharmacies, and submits complaints about the dodgy ones to the relevant authorities in the correct format. He has also taken to using another online tool to recruit and coordinate supporters who will make similar complaints, so that regulators such as the ASA and Trading Standards receive heavy correspondence on these issues. PledgeBank is a website that builds activist communities: people who sign up pledge to undertake a particular action if a certain number of others do the same. It's been used to ensure a copy of Richard Dawkins's *The God Delusion* is sent to every MP. Perry used it to enlist 115 people to make formal complaints to Trading Standards about 'Boots quackery'.

A similar purpose has inspired the launch of the Nightingale Collaboration, a voluntary group set up by two skeptical bloggers, Alan Henness and Maria McLachlan, to corral geek activism. It runs regular campaigns against specific aspects of alternative medicine, sifting the claims that are made for it and designing complaints to relevant authorities that are most likely to succeed. It also provides resources for those who want to do this themselves.

Another group of doctors and scientists used a different direct approach to target NHS funding of homeopathy and other alternative medicines. If the government won't insist on evidence, they reasoned, it might be possible to persuade the local bodies that make spending decisions: the Primary Care Trusts (PCTs). In 2006, Michael Baum and twelve senior medical and scientific colleagues wrote to every PCT in England demanding that they stop funding 'unproven and disproved treatments'.

The letter, and the media coverage it instigated, had a considerable impact. A year later, at least 86 of the 147 English PCTs had either ended or severely restricted funding for homeopathy; more than twenty of them had taken action in response to the letter. The doctors wrote again to the trusts to ask about progress and circulated a template document for evidence-based commissioning of homeopathy, which some began to use.

What made this initiative so successful was that it strengthened the positions of critics of alternative medicine working within the PCTs. 'As you would expect, within many of the organizations whose policies skip the evidence there are good scientists and doctors at work,' says Tracey Brown, director of the charity Sense About Science, who coordinated the letters. 'Their winces, grimaces and suggestions of an evidence-based approach can go unheeded amid political and bureaucratic imperatives. Until there is pressure from outside. A lack of evidence exposed in the press. There are questions in Parliament. Then the organizations start listening more to those people.'

The success of Baum's campaign, incidentally, may well now be threatened by the NHS reforms introduced in 2011. These involve abolishing the PCTs, many of which have become quite effective watchdogs on evidence-based spending, and replacing them with commissioning consortia in which GPs will have the dominant role. While it remains to be seen, at the time of writing, precisely how these will work, there is at least a danger that these consortia will be more easily influenced by individual GPs who advocate homeopathy, or by patient-led campaigns for more alternative medicine regardless of evidence. At minimum, there is a need here for renewed geek activism once the consortia begin work.

Geeks can also make a difference by exposing one of the few ways in which homeopathy can cause direct harm: when it is recommended in place of effective medicine for treatment or prevention of a serious disease. In 2006, Alice Tuff, of Sense About Science, teamed up with the BBC's *Newsnight* programme to pose as a young traveller to central Africa, where the most dangerous form of malaria is endemic. She visited ten homeopathic pharmacies, asking for advice about malaria prevention. All were prepared to prescribe homeopathic prophylactics, even though these are ineffective. None recommended that she take a conventional anti-malarial as well, and hardly any advised her how to prevent bites and recognize symptoms. One described a 'malaria-shaped hole in your aura'.

Even Peter Fisher, director of the Royal London Homeopathic Hospital, was shocked. 'I'm very angry about it because people are going to get malaria,' he told *Newsnight*. 'There is absolutely no reason to think that homeopathy works to prevent malaria and you won't find that in any textbook or journal of homeopathy so people will get malaria; people may even die of malaria if they follow this advice.'

Incredibly, none of the homeopaths Tuff exposed was disciplined by the General Pharmaceutical Council. But another Sense About Science activist, Julia Wilson, had more success with a campaign aimed at the World Health Organization (WHO). This UN body sets guidelines on the treatment and prevention of disease that are highly influential with national governments. Yet prior to 2009 it made no clear recommendations about the uselessness of homeopathy against conditions such as HIV, malaria, diarrhoea and tuberculosis. It was doing nothing to discourage people in developing countries – where these diseases are most prevalent and life-threatening – from turning to ineffective therapies.

Wilson organized an open letter to the WHO from twenty-five young doctors and scientists, many with experience in the developing world, which highlighted several examples where homeopathy was being promoted dangerously. It was sent to all the senior WHO officials responsible for these diseases, and Wilson followed up with calls and emails until she got a response. After three months, the WHO issued an unequivocal statement that it did not endorse homeopathy for any of these serious conditions, and many of its officials went further.

'WHO's evidence-based guidelines on treatment of tuberculosis have no place for homeopathic medicines,' said Dr Mukund Uplekar, of its TB Strategy and Health Systems Unit. The office of Elizabeth Mason, director of the Department of Child and Adolescent Health and Development, said: 'We have found no evidence to date that homeopathy would bring any benefit to the treatment of diarrhoea in children ... Homeopathy does not focus on the treatment and prevention of dehydration – in total contradiction with the scientific basis and our recommendations for the management of diarrhoea.'

Wilson distributed the WHO's new advice to health ministers all over the world. She ensured that an important message was communicated by a body to which governments would listen. 'Imagine how difficult it is for a junior doctor in a stressed region like East Africa, trying to get the authority figures in her local district not to listen to the material from a Dutch homeopathy clinic selling cheaper treatments than conventional medicine,' says Brown. 'Being able to point to WHO recommendations is really valuable.'

Beyond quackery

Homeopathy is the most egregious case in which the politics of healthcare matter more than the evidence. It certainly isn't the only one. Politicians waste their time on all manner of cranky health concerns that have little foundation in science because they are encouraged to take them up in the name of their constituents.

The alleged cancer risks from mobile phones, wifi networks and high-voltage power lines are a particularly attractive cause, because while there is little or no scientific evidence to indicate any danger, the siting of masts or introduction of wifi at schools are often hot-button constituency issues that bring people to MPs' surgeries. Four past or present MPs – Joe Benton (Labour), Ian Gibson (Labour), Willie Rennie (Lib Dem) and Caroline Lucas (Green) – serve as patrons of the EM Radiation Research Trust, which likens the cancer risks from mobile phones to that of asbestos and smoking. Tom Brake, a Lib Dem, secured a debate in 2007 at which he called for a ban on mobile-phone masts near schools and hospitals. Howard Stoate, a Labour MP who is also a GP, and Tim Loughton, a Tory, have agitated for a similar ban on building new homes near high-voltage power lines. Tessa Munt, a Lib Dem, has claimed, erroneously, that pylons have proven health effects.

This political attention stands out of all proportion to the risk involved, as there is neither epidemiological evidence of an epidemic of electromagnetically induced cancer nor a plausible physiological mechanism by which it might come about.

The perceived interests of constituents can also turn MPs into zealous opponents of evidence-based medical practice. Children's heart surgery is one of those highly specialized medical disciplines in which practice makes perfect. To achieve the best outcomes for their sick young patients, both surgeons and the units they work in need to conduct a high volume of operations so that their skills stay sharp. Diagnosis and treatment choices are also difficult to make, so it is important that surgeons have accomplished colleagues with whom to discuss patients and pool expertise.

The Department of Health therefore recommends that hospitals that operate on children's hearts should have four surgeons, and do 400–500 operations each year. Yet with about 3,500 operations in England each year, spread across eleven centres, the maths did not add up. Some of these units were going to have to merge to raise standards. In 2011, the Department proposed stopping cardiac surgery at four of these centres. These centres will remain open to see patients, but any who need operations will be sent to the larger, merged ones.

A merger between two units, however, can easily be portrayed as the closure of one. And the closure of any local hospital service can be guaranteed to pique the campaigning instincts of the local MP. Never mind that the decision might be medically justified and in the best interests of the people being served; it is just the sort of emotive issue around which a reputation for fighting for constituents can be made.

MPs representing the downgraded hospitals duly kicked up an almighty fuss. 'Intellectually, the case for change is compelling,' said Sir Bruce Keogh, the NHS Medical Director. 'Sadly, the realpolitik is that the closer we get to a solution, the more personal, professional and political interests conspire to perpetuate mediocrity and inhibit the pursuit of excellence.'

Guided by the science?

When science and medical ethics mix in the political arena, there is still greater potential for evidence to be ignored and even abused. Abortion is not an issue that can be decided by evidence alone – science cannot ultimately determine whether it is right or wrong to allow women to terminate an unwanted pregnancy. Science can, however, establish much about the viability and capacity for pain of foetuses at different stages of gestation. As many people who do not take an absolutist pro-life or pro-choice position think these factors relevant to setting appropriate time limits, politicians often bend the evidence to support their case.

Nadine Dorries, a Conservative MP who has made restricting access to abortion her signature political cause, is a particular offender. Dorries has claimed time and again that her support for a lower limit is based not on her religious faith, but on science. The twenty-four-week limit is too high, she argues, because of growing evidence that very premature infants born earlier than this are increasingly likely to survive and that foetuses at this stage can feel pain. She also proposed an amendment to health legislation in 2011 that would have made it mandatory for women considering an abortion to be offered independent counselling about the physical and psychiatric risks.

Following the science, however, is not helpful to Dorries's cause. In 2010, the Royal College of Obstetricians and Gynaecologists published an extensive review of foetal pain, which concluded that the nerve connections to the cortex that are necessary for pain perception do not form until after twenty-four weeks. It is a similar story for pre-term survival. The best evidence comes from two separate studies, the Trent study and the EpiCure2 study, which agree that while survival at twenty-four and twenty-five weeks is improving rapidly, the same is not true for infants born at twenty-two and twenty-three weeks.

Abortion can, of course, lead to complications such as infection and uterine perforation, though these are rare, and there is evidence that women who have them have a slightly raised risk of miscarriage and pre-term delivery in subsequent pregnancies. But doctors are already required to apprise women of these risks, under evidence-based guidelines for obtaining informed consent.

In her eagerness to justify her counselling amendment, Dorries quoted scientific evidence in highly selective fashion. In her Commons speech proposing her counselling plan, she highlighted a single study suggesting that 'women who have an abortion are twice as likely to suffer from mental health problems'. She failed to refer to a more exhaustive American Psychological Association report, which concluded that adult women with an unplanned pregnancy who have a first-trimester abortion are at no more risk of mental illness than those who have the baby. Also overlooked was the draft conclusion of a Royal College of Psychiatrists review – available when she spoke – that 'mental health outcomes are likely to be the same, whether women with unwanted pregnancies opt for an abortion or birth'.

As her Tory colleague Sarah Wollaston – a GP with a much better grasp of evidence-based medicine – pointed out in the debate: 'My honourable friend has twice quoted the Royal College of Psychiatrists and asserted that there is a much higher rate of mental illness after termination of pregnancy, but the RCP has made it clear – any member can look online at the draft of its very comprehensive evidence review – that we have to compare like with like. In other words, we have to make a comparison with rates of mental illness after unwanted pregnancy. Looking at the rates after unwanted pregnancy, we see that there is no difference between the rate of mental illness after termination of pregnancy and live birth. Indeed, the biggest predictor of mental ill health after a termination of pregnancy is whether somebody was suffering with problems beforehand.'

Dorries's counselling proposal was designed to sound reasonable, but it would at best have tackled a problem that didn't really exist. It's hard to escape the conclusion that it was really intended to slow access to abortions and to allow anti-abortion groups to counsel women about harms that aren't supported by evidence. Thankfully, it was voted down – though it's somewhat alarming that she found another 117 MPs to support it.

Dorries also loves to make her case with a good anecdote, most notably when she used a photograph nicknamed the 'Hand of Hope', which purported to show Samuel Armas, then a foetus of twenty-one weeks, reaching out to grab the finger of a surgeon performing an intra-uterine operation. Ben Goldacre has exposed this as a hoax. Joseph Bruner, the surgeon, explained: 'Depending on your political point of view, this is either Samuel Armas reaching out of the uterus and touching the finger of a fellow human, or it's me pulling his hand out of the uterus . . . which is what I did.'

Anti-abortion campaigners such as Dorries are of course perfectly entitled to their viewpoint: where you stand on abortion is ultimately a matter of personal ethics, not science. But if you're going to use science, rather than religion or personal ethics, to make your case, you have a duty to get the science right. Dorries's distortions matter because they have often been taken at face value by other MPs who have engaged less directly with the evidence. David Cameron, then leader of the Opposition, said that he would be guided by the science when he voted on revising the abortion time limit in 2007. He supported a cut to twenty-two weeks.

This sort of science abuse is particularly rife in the United States, where conservative Republicans and religious groups have routinely used it to justify policy prescriptions that would be more honestly described as faith-based. President George W. Bush's Administration was guilty of many such distortions, as Chris Mooney details in *The Republican War on Science*. The US National Cancer Institute website was edited to suggest a link between abortion and breast cancer that does not exist. The Food and Drug Administration overruled its scientific advisers' recommendation to make the morning-after pill (known as Plan B in the US) available over the counter. The Centers for Disease Control and Prevention withdrew a list of sex-education programmes that had been shown to be effective, while there was heavy investment in unproven abstinence-only programmes.

Immune to evidence

Even a stopped clock tells the right time twice a day, and sometimes even serial political science abusers get it right. On climate change and evolution, Rick Perry, the Republican Governor of Texas and 2012 presidential candidate, is about as wrong as they come. In 2007, however, he took a courageous and correct decision about public health. Perry signed an executive order that provided for eleven- and twelve-year-old girls in Texas to be vaccinated against HPV, the virus that causes most cases of cervical cancer, provided their parents did not object.

Gardasil, the vaccine Perry endorsed, is highly effective at preventing infection with HPV. His decision, though, proved controversial. It angered other religious conservatives who felt a vaccine against a sexually transmitted virus might promote pre-marital sex – as if libidinous teenagers stop to think about cervical cancer. It upset right-wingers who object to state-sanctioned vaccination programmes as an infringement of personal and parental liberty. And it raised questions about Perry's links with Merck, the vaccine's manufacturer, who had contributed to his campaign. The state legislature overturned the order.

This evidence-free row means that hundreds of Texan women will needlessly develop cervical cancer. Its impact, too, is now going to be felt well beyond Texas, because of the way it has been cynically exploited by one of Perry's political rivals.

At a Republican primary candidates' debate in September 2011, Michele Bachmann, the Minnesota congresswoman, accused Perry of putting young girls' health at risk to reward a drug company. She followed up with wild claims that the HPV vaccine was dangerous. 'I will tell you that I had a mother last night come up to me here in Tampa, Florida, after the debate,' Bachmann told NBC's *Today* show. 'She told me that her little daughter took that vaccine, that injection, and she suffered from mental retardation thereafter. It can have very dangerous side effects.'

Bachmann's assertion was unfounded. There is no evidence at all that HPV vaccines cause mental retardation – in fact, they have an excellent safety profile. She promoted a groundless scare about a life-saving immunization to score political points. If substantial numbers of parents listen to this sadly influential political leader, and refuse to allow their daughters to be vaccinated, there are going to be needless deaths. It is hard to think of a clearer case in which bad science has crossed the line into dangerous irresponsibility.

Bachmann's vaccine scaremongering is unusual among politicians. Most support immunization programmes, and many have been remarkably resilient in that support, given the many baseless vaccine scares that have emerged in recent years. When Andrew Wakefield made his disproved claims of a link between MMR and autism, the UK government stood firm in the face of often overwhelming media pressure to replace the triple jab with separate immunizations against measles, mumps and rubella.

This was the proper evidence-based decision: there was no evidence that MMR carried any more risks than those attached to any vaccine, while single jabs, requiring six visits to a GP in place of the two needed for MMR, could have significantly reduced coverage, particularly among vulnerable and low-income groups. Ministers deserve credit for heeding doctors and scientists who knew what they were talking about over shrill newspaper columnists and a handful of MPs from the Bachmann tendency.

Vaccine scares, as it happens, show a curious cultural specificity. In the UK, the MMR vaccine has been the main target of suspicion. In the US, it has been vaccines containing the mercury preservative thimerosal; and in France, jabs against hepatitis B and multiple sclerosis. In the UK, the *Daily Mail* campaigns against the HPV jab, which the government supports. In Ireland, where the government has refused to introduce it, the same newspaper campaigns in favour of the same vaccine. It is the decisions made by journalists, grand-standing doctors and ignorant politicians that foment vaccine panic, not scientific evidence of risk.

Politicians can contribute with sins of omission as well as com-mission. In May 2000, Tony Blair became the first serving prime minister in more than 150 years to father a legitimate child while in Downing Street, and as the MMR controversy raged, his press office began to field questions from reporters about whether or not the baby boy, Leo, had had the vaccine. Blair resolutely refused to answer, stating again and again that his family's health was an entirely private matter that he was not obliged to share with the press.

Blair's response was understandable on one level: medical matters are rightly considered confidential, and it is easy to see why he and his wife, Cherie, were wary of setting a precedent. That said, all he was being asked was to confirm that his son had had a normal vac-cination that his government was recommending for every child of similar age in the country. It was not as if the media were seeking to expose a rare and distressing disease.

The Blair family fitted the profile of MMR refuseniks. Cherie Blair is well known as an enthusiastic user and supporter of the kookier end of alternative medicine, with a penchant for crystals and dowsing. It wasn't unreasonable for the press to imagine that Leo might not have been given the vaccine.

Blair's refusal to address the issue presented the MMR vaccine's media opponents with an opportunity to turn the scare into a proper political issue. It put MMR back on the front pages, amid innuendo that the Prime Minister, or his wife, did not trust the official advice being given out by his own Health Secretary. Blair's silence was perplexing because, he later admitted in his autobiography, Leo Blair had had the vaccine. A simple statement would have defused an issue that did grave damage to public health: MMR coverage fell from 88 per cent in 1988–9 to 80 per cent in 2003–4. Blair had an opportunity, a responsibility even, to lead by example on this issue, and he shirked it. As Michael Specter, the American science writer, wrote of the issue: 'No virus respects privacy . . . so public health is never solely personal, as the impact on Britain has shown.'

Vaccine scares such as the MMR panic also highlight the important role that individuals can play in protecting public health. While Andrew Wakefield's research into MMR and autism was widely criticized from the outset, it was the dogged investigative work of Brian Deer, a freelance journalist working principally for the *Sunday Times*, which eventually uncovered his extensive conflicts of interest and unethical research practices. The evidence that Deer pieced together was eventually sufficient to secure the retraction of the *Lancet* paper that began the scare, and to convict Wakefield of serious professional misconduct by the General Medical Council, which struck him off the medical register. Without Deer's forensic investigation of this enemy of public health, the damage Wakefield caused would arguably have gone further.

As online tools have made it ever simpler to publish, geeks have unprecedented opportunities to investigate similar health scares and to bring their instigators to book. There's also much we can achieve at a more informal level. First, we can ensure that we offer support to the scientists, doctors, journalists and other experts who stand up for the importance of evidence-based medicine in public life, those who defend vaccines against scaremongers like Wakefield and Jenny McCarthy, the *Playboy* model who leads the anti-vaccine campaign in the US. The people who stand up to them – such as Deer, David Elliman of Great Ormond Street Hospital, and Paul Offit of the University of Pennsylvania – often find themselves vilified. All three have been targeted with extreme personal abuse, accused of being pharmaceutical industry 'shills' or worse. Entering this arena is not something that can easily be done without a thick skin, but those who do it need to be told that what they are doing is valued.

We can also set an example in our personal lives, by explaining the science behind vaccination, or homeopathy, to our own friends, family members and work colleagues. There is always a temptation to stay silent in the face of ignorant claims about health, whether anecdotal recommendations of alternative medicines or ill-informed attacks on vaccine safety. These are subjects over which it is easy to start a fight. But it's important that we challenge these beliefs when we can. We need to explain to the people we know why we have our children vaccinated, why the very real risk of measles outweighs the illusory one of autism, and that using homeopathy will at best waste your money and at worst damage your health if it's used in place of a real drug that's needed.

Some geeks are doing this already, taking the initiative to challenge the prevailing discourse over healthcare by insisting that evidence matters. How much more might we achieve if everybody who values the scientific method and evidence-based medicine joined the activist ranks. Alan Henness and Maria McLachlan hit the right note when they started the Nightingale Collaboration. 'Misleading information won't disappear by itself,' they said. 'It needs to be challenged.'

That isn't just true for medicine. The politics of the environment are replete with misleading information, and its purveyors come from more than one side.

GEEKS AND GREENS

Why science matters to the environment

SOON AFTER GEORGE W. BUSH BECAME US PRESIDENT IN 2001, THE Republican pollster Frank Luntz sent a confidential memo to the party's leadership. The environment, he told them, was 'probably the single issue on which Republicans in general – and President Bush in particular – are most vulnerable'. A growing scientific consensus on global warming was threatening to cast the party as 'in the pockets of corporate fat cats who rub their hands together and chuckle manically as they plot to pollute America for fun and profit'.

Luntz's message was not that the Republicans should listen to what the majority of scientists were telling them about climate change, to detoxify their environmental image by following the evidence. It was that public perceptions of the science could still be changed to fit the pre-existing Republican rhetoric, given the right sort of campaigning.

'The scientific debate is closing [against us] but not yet closed,' Luntz wrote. 'There is still a window of opportunity to challenge the science. You need to be even more active in recruiting experts who are sympathetic to your view, and much more active in making them part of your message. People are willing to trust scientists, engineers, and other leading research professionals, and less willing to trust politicians.'

It was a message that the Republicans took firmly to heart. As the House of Representatives Oversight and Government Reform Committee concluded in 2007, the Bush Administration 'engaged in a systematic effort to manipulate climate change science and

214 THE GEEK MANIFESTO

mislead policymakers and the public about the dangers of global warming'.

White House officials repeatedly edited reports from the Environmental Protection Agency to play down scientific confidence about the existence of climate change, the role of human activity in its origins, and its likely impacts. The agency was also pressed to refuse to regulate greenhouse gases as pollutants.

Scientists employed by the government with inconvenient views were muzzled, as the White House insisted on approving media requests for interviews. Even congressional testimony was altered. When Thomas Karl, director of the National Climate Data Center, gave written evidence in 2006, he was forbidden from saying that 'modern climate change is dominated by human influences', that 'we are venturing into the unknown territory with changes in climate', or that 'it is very likely (>95 per cent probability) that humans are largely responsible for many of the observed changes in climate'.

The strategy also exploited the media's tell-both-sides model of political reporting that serves science so badly, using contrarian scientists such as Fred Singer, a retired physicist, to create the misleading impression that the science of climate change was riven with doubt and dissension.

This cynicism successfully kept Luntz's window of opportunity ajar. Then, immediately before world leaders gathered in Copenhagen at the end of 2009 to discuss international action on climate change, the global-warming deniers were handed a crowbar.

In November 2009, a computer hacker broke into a server at the University of East Anglia's Climate Research Unit and stole more than 1,000 emails sent by and to its scientists, notably the centre's director, Professor Phil Jones. On 19 November, the hacker, or someone working with him, uploaded the entire cache of documents to a server in Tomsk, Russia, and it was rapidly copied to multiple websites. Then the climate-change contrarians at blogs such as *Watts Up With That* and *The Air Vent* got to work.

James Delingpole, a right-wing journalist who blogs for the *Daily Telegraph*, coined the phrase that stuck. 'Climategate', as he called it, could be 'the final nail in the coffin of anthropogenic global warming'.

To these deniers and contrarians, the UEA emails revealed a massive scientific conspiracy to misrepresent climate research, to manipulate data, to silence critics and to cover up inconvenient findings. They took quote after quote from leading scientists out of context to suggest that even 'warmist' scientists, to use their pejorative language, knew that the case for man-made global warming was fundamentally flawed.

Taken in isolation, some of the emails looked damning. 'The fact is that we can't account for the lack of warming at the moment and it is a travesty that we can't,' wrote Kevin Trenberth, of the US National Center for Atmospheric Research, in one email. Professor Jones wrote of using a 'trick' to 'hide the decline', and boasted in another email that he and Trenberth would keep a paper questioning global warming out of the latest Intergovernmental Panel on Climate Change (IPCC) review of the science, 'even if we have to redefine what the peer-review literature is!'

Other emails reflected badly on the scientists in different ways. Jones described the death of a sceptic named John Daly as 'cheering news', and appeared to advocate deleting data so it could not be released under the Freedom of Information Act. The overall impression could scarcely have been more damaging. At best, the scientists who sent the emails looked vindictive, secretive and defensive. At worst, they found themselves accused of conspiring to commit scientific fraud. Right-wing politicians such as James Inhofe, Sarah Palin, Michele Bachmann and Rick Perry seized on the emails to pronounce global warming a gigantic hoax.

The damaging revelations didn't finish with the emails. In January 2010, Fred Pearce of *New Scientist* revealed a significant error in the IPCC's 2007 report. The expert panel's claim that all the glaciers in the Himalayas could melt by 2035 turned out to have been based on a speculative, throwaway remark by an Indian scientist in a magazine interview, which found its way into a 2005 report from the World Wildlife Fund, an environmental lobby group. This was the signal for journalists to start combing the thousand-page report for further flaws. A few were found, such as an assertion that 55 per cent of the Netherlands lies beneath sea level, when the true figure is 26 per cent. The wrong information had been supplied by the Dutch government, but it was an inexcusable error of peer-review.

It is difficult to say how public opinion was affected, because of a shortage of tracking polls that asked precisely the same questions before and after the furore. There is at least some evidence of an impact. Polls conducted by Populus before and after the controversy found that the proportion of British adults who did not believe climate change was happening rose from 15 per cent in November 2009 to 25 per cent in February 2010. The proportion agreeing that climate change is definitely happening and proven to be man-made fell over the same period, from 41 per cent to 26 per cent. The proportion that thinks climate change to be environmental propaganda, however, was unchanged – a surprising result if Climategate were driving the effect – and other poll comparisons suggest there has been no great lasting effect.

What Climategate certainly changed, though, was the media narrative. At the very moment when world leaders were discussing how to respond to climate change, the focus shifted to whether it was happening, and whether scientists could be trusted. Conservative newspapers that had softened sceptical coverage of global warming, such as the *Daily Mail*, became emboldened and more hostile. The BBC began to bend over backwards to balance scientific opinion with critics' counter-claims, often using the Global Warming Policy Foundation, a new contrarian think-tank founded just as the controversy broke.

Climategate turned climate denial into a respectable and mainstream position again. It gave the Luntz strategy the impetus it needed to succeed.

Luntz, incidentally, has disowned his own advice, as he became convinced by the science he had once sought to undermine. But that has not stopped it from becoming something close to an article of faith among Republicans. Mitt Romney's broad acceptance of the science was widely perceived as a handicap in his campaign for the Republican presidential nomination in 2012, against denialist rivals such as Bachmann and Perry – so much so that he has begun to backtrack.

Yet Climategate and the IPCC disclosures that followed have done nothing to undermine the consensus scientific position on global warming. Even after peer-review, a document as extensive as the IPCC report was always likely to contain a few errors. These are regrettable, but none casts any doubt on its central conclusions about the human origins of observed climate change, or its likely progress over the next century. Neither did the UEA emails, which contained far less of substance than initially met the eye.

They revealed no evidence that scientists had participated in any sort of scientific misconduct or fraud. The apparently damning quotations are damning only when taken out of context. Trenberth's comment about accounting for a lack of warming referred to a specific problem in calculating energy flows over the past decade, not to the overall thrust of climate change. The 'trick' Jones alluded to was a statistical technique used to combine instrumental and tree ring data, and the 'decline' was an odd fall in tree growth at a time when observed temperatures were rising. There was no deception to cover up an inconvenient cooling trend. As the leading journal *Nature* put it in an editorial: 'A fair reading of the emails reveals nothing to support the denialists' conspiracy theories.'

Even had the emails called the integrity of the UEA's global temperature data into question, this research was not the sole source of data in the field. Two separate American datasets, from Nasa and the National Oceanic and Atmospheric Administration, showed broadly similar patterns of warming. In the months that followed the theft, independent inquiries ordered by UEA, the Royal Society and the Commons Science and Technology Committee duly cleared the scientists of research malpractice, and pronounced the underlying science of global warming unaffected. Expert bodies such as the American Meteorological Society and the American Geophysical Union have reviewed their positions in light of the controversy, and altered nothing.

If the saga revealed anything about climate science, it is the lengths to which contrarians will go to cherry-pick arguments that appear to support their point of view.

Nevertheless, UEA and its scientists cannot escape all blame for the public-relations disaster. While the unguarded email correspondence did not reveal any misconduct or figure-fiddling, it did expose what the Commons committee called a 'culture of non-disclosure'. It is obviously tempting for researchers to refuse to help critics seeking to pick holes in their work, but open data is an important principle of science, which allows research to be checked by anybody who wishes to do so. In some other fields, such as genomics, it is widely accepted that raw data should be released as it is gathered, to create a public resource. A more open approach would have shot the fox of deniers who used climate science's lack of transparency to insinuate that it had something to hide.

These insinuations were also allowed to fester by the inadequate response of the protagonists when the emails were published. For two weeks, UEA maintained what it thought was a dignified silence. Rather than address the content of the emails and explain the context in which apparently incriminating remarks had been made, its line was that it would not discuss documents that had been stolen. Never likely to succeed, this tactic was all the more remarkable as Climategate erupted just months after the theft of a computer disc had revealed full details of dubious expense claims made by dozens of British MPs. The provenance of that information did nothing to save politicians from public ire, any more than it was likely to protect the scientists.

No robust defence of Phil Jones and his colleagues was offered, and the scientists went to ground. The result was that the climate sceptics' claims went unchallenged and accordingly framed the way in which the whole incident was portrayed to the public. As uncontested allegations against Jones gained more and more currency in the press, even scientists and green campaigners hesitated to speak up for him. George Monbiot, the influential green activist and *Guardian* columnist, called for Jones to resign. As the journalist Fred Pearce noted, potential sympathizers became 'unwilling to defend people whose employers were leaving them to hang in the breeze'.

The climate scientists and UEA took the view that their science was sound and that the processes of science would ultimately bear this out. On one level, they were right: the investigations that followed cleared them of any wrongdoing, and there has been no impact in the peer-reviewed literature. Yet on another level, they missed the point. Public perceptions weren't going to be moulded in the long run, but here and now. The situation called for the accused scientists and their allies to meet their foes in hand-to-hand combat. Instead, they retreated to the citadel of academia, leaving the enemy free to pillage beyond the redoubt.

This problematic attitude was encapsulated by a remark Jones made to Sir Paul Nurse, the Nobel laureate and new President of the Royal Society, during a BBC *Horizon* documentary broadcast a year on from the crisis. 'Much of the science is in the peer-reviewed literature,' he said, 'and I wish people would read that instead of the emails.' It's an understandable attitude for scientists to take when their work is attacked, but a wholly unrealistic one. Most people have neither the technical skill nor the inclination to read the peer-reviewed literature – they will rely on others to sum it up. If the scientists who can best explain it stay silent, then those who wish to misinterpret it to their advantage will get a clear run.

Jones himself can perhaps be excused for failing properly to defend his work. Climategate put him under such intense personal pressure that he rapidly lost weight and even contemplated suicide. But his colleagues and university chiefs should have engaged on his behalf. It wasn't as if the case was difficult to make. On *Horizon*, Jones gave Nurse a crystal-clear and convincing explanation of his 'trick' comment. How sad that it came a year too late.

Scientists can't afford to let their science speak for itself: they have to stand up for their work and their integrity, and make a case that the public can follow. As the description for a session on Climategate at the Science Online conference in 2011 put it: 'For many scientists, fighting back means publishing a really good paper in a reputable journal. That doesn't cut it any more.'

'For many scientists, fighting back means publishing a really good paper in a reputable journal. That doesn't cut it any more' — Science Online Conference, 2011

The perils of opportunism

If incompetent and defensive PR was a proximate cause of the crisis in climate science that the hacked emails were allowed to provoke, its roots ran deeper. The tactics adopted by some climate scientists, and particularly by their champions in the green movement, had at once invited some of the unfounded charges advanced by the deniers and made them harder to defend.

Many of those who rightly insist that the science of climate change is well established have also been guilty, on occasion, of mis-representing it to serve their purposes. The result has been to undermine what ought to be a compelling case by leaving the public in doubt about who to trust.

Over the past two decades, as the scientific consensus on global warming has become ever clearer, groups such as Greenpeace and Friends of the Earth have campaigned with great success to raise its political profile. Brilliant PR machines, allied to astute lobbying operations, have turned climate change and the environment into mainstream political issues that no MP, congressman or senator can now ignore. That the UK is now legally obliged to cut carbon emissions by 80 per cent by 2050 is in large part down to Friends of the Earth, which mobilized hundreds of thousands of supporters to write to MPs ahead of the 2008 Climate Change Act. The greens' achievement is in many ways something which geeks should be seeking to emulate if we're to embed evidence and scientific thinking more deeply in the political process.

Yet the green approach has also demonstrated the pitfalls of playing fast and loose with the evidence. The core of its message on climate change is that science tells us that global warming is happening, that it is potentially dangerous, and that human activities are at least partially responsible. Each of these propositions is well founded. Green campaigning on the issue, however, often goes further, stepping beyond the bounds of scientific knowledge and uncertainty for the sake of building a more colourful case.

Extreme weather is a good example. While computer projections of how a warmer world will look suggest that there will be more storms and floods, heatwaves and droughts, most climate scientists are extremely wary of attributing any particular weather event to global warming. Hurricanes and forest fires may become more frequent if temperatures continue to rise, but they have always happened, always will happen, and aren't necessarily caused by climate change. That caveat is often ignored by green campaigners, for the simple reason that linking the natural disaster that has hit the headlines to global warming makes for a powerful narrative that appears to help their case.

When Hurricane Katrina hit New Orleans in 2005, Greenpeace called it 'a wake-up call about the dangers of continued global fossil fuel dependency'. Al Gore suggested that warmed Caribbean waters may have made the storm stronger, adding to the devastation it caused. There is some evidence that global warming may increase the intensity and frequency of hurricanes, but it is impossible to say whether it played a role in Katrina. Similar claims were made in Australia when severe wildfires struck Victoria in 2009, and when Queensland experienced catastrophic floods in 2011. The 2003 European heatwave that claimed 40,000 lives, the autumn flooding that hit the UK in 2000, the extreme tornadoes that struck the US in 2011: all have been claimed as evidence that climate change is already happening.

Each of these events is consistent with global warming. Many of them will become more frequent in a warmer world: Peter Stott, of the UK's Met Office, and Daithi Stone and Myles Allen, of the University of Oxford, have shown it is highly likely that anthropogenic emissions have doubled the risk of heatwaves on the scale of 2003.

But there is a fine line between using this sound statistical argument to illustrate what climate change might look like and suggesting that particular weather events prove that the negative impacts of global warming are already with us. By claiming them as direct evidence for the prosecution, greens have been playing a very dangerous game.

First, this line of argument is easily rebutted. It creates an unnecessary weakness which deniers can target to sow doubt about the rest of the science. If Al Gore and Greenpeace erroneously claim the weather in evidence, what price the accuracy of the rest of their supporting data? Every time a campaigner exaggerates the evidence for climate change, he risks later being hoist by his own petard.

What is more, if a weather event that is consistent with climate change can be used to make a point, so can one that appears to contradict warming. Thus every heavy snowfall, every unseasonable cold snap, every year without an increase in global temperatures becomes an opportunity to question whether climate change is real. Extreme weather doesn't refute global warming any more than it confirms it – it is long-term trends that matter – but public perceptions are easily influenced this way.

As snow fell heavily in Britain in December 2010, climate sceptics began to forward one another links to an *Independent* headline from 2000, which proclaimed that 'snowfalls are now just a thing of the past'. It went viral, as it seemed elegantly to show that extreme climate predictions can rapidly be proved wrong. Leo Basari, who writes the *Climate Sock* blog about public opinion of climate science, has even suggested that Britain's cold winter of 2009–10 did more to feed global-warming scepticism than Climategate.

A Friends of the Earth press release from January 2011, in response to new data placing 2010 as the joint second warmest year on record, illustrates the contortions that the weather narrative can require. In one breath, Craig Bennett, the group's policy and campaigns director, was explaining, reasonably and correctly, that: 'We may have been trudging through deep snow last month, but these figures show there's an enormous difference between climate and weather – despite the efforts of some to claim that global warming is a myth.' Yet in the very next paragraph, Bennett went on to make the same mistake: 'Experts say climate change will increase extreme weather events, and from Brisbane to Brazil there have been numerous weather-related disasters in recent weeks.'

Is it surprising that many people wonder whether the scientific case for global warming might have been oversold, when its most prominent proponents want to have their cake and eat it?

This dangerous inconsistency isn't restricted to green PR. The movement's rhetoric on climate change always presses home the strength of the scientific consensus. But when an expert consensus happens not to suit its agenda, it is more than happy to ignore it.

A radioactive issue

Shortly after lunchtime on Friday, 11 March 2011, a geological fault 70 kilometres off the east coast of Japan ruptured, triggering a magnitude 9 earthquake. The Tohoku tremor, the joint fourth most powerful on record, launched a series of huge tsunami waves which began to wash ashore a few minutes later. The destruction, principally wrought by the tsunami, was phenomenal: more than 15,000 confirmed deaths, more than 4,000 still missing in August, five months later, and 125,000 buildings destroyed or damaged.

At the Fukushima Daiichi nuclear power plant, about 120 kilometres (75 miles) from the epicentre, three out of six reactors were operating when the quake struck. Initially, they coped pretty well. The shaking triggered an emergency shutdown of the main fission reactions and, though the main electricity supply was cut off, auxiliary generators kicked in immediately so that cooling water continued to be pumped through the cores. Three quarters of an hour later, the tsunami arrived. Waves 14 metres high overwhelmed a sea wall designed to cope with just a 5.7 metre tsunami, and the generators, sited at ground level, were knocked out. When back-up battery power ran out too, pumps could no longer circulate cooling water and the reactors began to go into meltdown.

The world's worst nuclear disaster since Chernobyl had begun.

As residents were evacuated from an exclusion zone 20 kilometres around the Fukushima plant, the situation escalated into a crisis. Explosions ripped through the outer buildings housing several of the reactors, as hydrogen produced by chemical reactions within the overheated cores was vented to reduce pressure, then ignited as it accumulated (some of the media disingenuously described these as nuclear explosions). Rising radiation readings led Tepco, the power company that operates the complex, to evacuate all but fifty key workers.

Sea water was pumped into the reactors as an emergency cooling measure that would wreck them for good. As pools full of spent nuclear fuel rods also began to overheat, helicopters and water cannon were deployed to douse them. While there was no fire or explosive breach of containment as happened at Chernobyl, large amounts of the radioactive isotopes iodine-131 and caesium-137 were released, principally into the sea. Concern about the spread of radiation was such that residents of Tokyo, 240 kilometres (150 miles) away, were for a time advised to avoid giving tapwater to infants, and several countries, including France, evacuated their nationals from the capital.

Fukushima was an exceptionally serious nuclear incident, and the bravery of the 'Fukushima 50's' struggle to contain it made for a compelling news story. But the alacrity with which it was misleadingly exploited by the green lobby to campaign against nuclear power was none the less breathtaking. It was an episode that highlights the movement's habit of taking or leaving mainstream scientific opinion according to its suitability to its purposes.

As a low-carbon means of generating electricity with a good safety profile, nuclear power is a technology with widespread support among scientists and engineers. Britain's Royal Society and Royal Academy of Engineering both regard it as essential to containing climate change, and a 2009 consensus statement from the science academies of thirteen countries recommended the development of 'safe and secure nuclear power capacity' as a policy priority.

Yet despite nuclear energy's manifest potential to contribute to the world's greatest environmental challenge, all the leading environmental groups, including Greenpeace, Friends of the Earth and the Green Party, stand resolutely against it. As far as they are concerned, the health consequences of nuclear accidents, and the problem of disposing of radioactive waste, put nuclear power beyond the pale – whatever science might have to say to the contrary.

As soon as it became clear that the Fukushima plant was in trouble, Greenpeace dispatched activists in white protection suits, brandishing Geiger counters, to make the most of a photo opportunity. 'How many more warnings do we need before we finally grasp that nuclear reactors are inherently hazardous?' asked Jan Beranek, head of nuclear campaigns for Greenpeace International, on the day after the earthquake and tsunami.

All over the world, anti-nuclear campaigners, spearheaded by Green parties and the main environmental NGOs, followed Beranek's lead, using Fukushima to put pressure on politicians to abandon nuclear power. In several places they succeeded. Germany shut down seven of its seventeen reactors immediately and will phase out the rest by 2022. Italian voters overturned a proposal to build new nuclear plants in a referendum, and Switzerland decided not to replace its five plants when they reach the end of their lives. Public opinion has shifted against nuclear in many nations. A twenty-four-country poll by Ipsos-MORI in May 2011 found that 62 per cent were opposed to nuclear power, with 26 per cent saying that Fukushima had influenced their position.

Most worryingly, given its hunger for energy, China experienced a nuclear panic, amid unfounded fears that a radioactive cloud was drifting west. Supermarkets ran out of salt, which was erroneously thought to protect against radiation. The Chinese government suspended the approval of new reactors.

This international crisis of nuclear confidence, stirred up by the greens, spells extremely bad news for the climate.

If global warming is to be contained to manageable levels, there is growing agreement among climate scientists that the concentration of carbon dioxide in the atmosphere needs to be stabilized at no more than 350 parts per million (ppm). That should keep global temperature increases below about 2°C and prevent some of the worse potential consequences.

This target can be achieved only by decarbonizing the world's energy systems, and as the 350ppm threshold was crossed in 1988, and atmospheric carbon dioxide currently stands at about 390ppm, the urgency of the situation is plain. An expansion of renewable sources of energy, such as wind and solar, is going to be necessary if we're to get there, but it won't, by itself, be enough. We also need another source of low-carbon power that is already mature enough to provide a significant proportion of our base-load energy needs.

That source should be nuclear power. The world's installed nuclear capacity already generates 15 per cent of electricity, preventing the emission of over 2 billion tonnes of carbon dioxide each year that would otherwise be released by fossil-fuel plants. China, which overtook the US in 2007 as the world's biggest source of greenhouse emissions, is currently building twenty-six nuclear plants and plans to generate 70 to 80 gigawatts of electricity (GWe) from nuclear power by 2020, 200 GWe by 2030 and 400–500 GWe by 2050. If green activists succeed in derailing this low-carbon programme, coal-fired plants will be built instead, at immense environmental cost: coal releases about 6 million tonnes more carbon a year than nuclear per gigawatts of electricity.

Germany has the long-term aim of generating all its electricity renewably, but in the short term its scrapped nuclear plants aren't going to be replaced with wind turbines or solar panels, but with old-fashioned coal. As a result, it will emit 300 million tonnes more carbon than it would have done had it not said 'Atomkraft? Nein danke.' If China follows suit, the impact on the climate will be greater by orders of magnitude. As Mark Lynas, an environmental campaigner and journalist, argues in his recent book The God Species, the green movement's hostility to nuclear power has already worsened climate change by blocking more widespread adoption in the 1970s and 1980s, when parts of the United States and Austria, among other countries, went for coal instead. It is now using Fukushima to repeat the mistake.

It is a mistake that is founded on a fiction. For far from being 'inherently hazardous', as Greenpeace claims, nuclear power is in fact considerably safer than its principal alternatives, even leaving climate impacts aside. Fukushima was the world's worst nuclear disaster for a quarter of a century. Yet at the time of writing, six months after the accident, the death toll from radiation stands at nil.

In the same month, a coal-mine explosion in Baluchistan province, Pakistan, killed forty-five workers. Another forty-eight American coalminers died at work in 2010; its nuclear industry had no fatalities. China's coal industry claimed 2,631 miners' lives in 2009 – the year with the best safety record in the past decade. In September 2011, four miners died when a mine flooded in Wales. Coal, which is what gets used today when nuclear power is un- available, has been estimated to cause 161 deaths per terawatt hour (TWh) of energy, compared to 0.04 deaths for nuclear. On this measure, nuclear is safer even than wind power, which has a death rate of 0.15 / TWh.

The health hazards of radiation released by nuclear accidents have been systematically over-egged by green campaigners. Chernobyl is often claimed as evidence of the technology's calamitous potential: Greenpeace asserts that it may cause 100,000 cancer deaths. Yet that accident, far more serious than Fukushima and, as we will see, an exceptional one, has actually caused about sixty deaths in twenty- five years. UNSCEAR, the United Nations committee that continues to evaluate Chernobyl's health effects, states that there were 134 cases of radiation sickness among the 'liquidators' who fought the reactor fire, twenty-eight of which were fatal within three months. Another nineteen died between 1987 and 2006, though it is unknown how many of these deaths were related to radiation; at least one, from trauma, was not.

About 6,000 children and adolescents have contracted thyroid tumours, though as this cancer is treatable, only fifteen cases were fatal. Most – perhaps even all – of these cases could have been pre- vented had the Soviet authorities distributed iodine tablets, as the Japanese did after Fukushima. Though the UN once estimated that, in the worst case, up to 9,000 future cases of cancer may be attributable to radiation from the accident, its most recent reports have found no evidence for higher rates of leukaemia in anybody other than liquidators, and no evidence in any group for a raised incidence of solid tumours other than thyroid cancers.

Fukushima's consequences have yet fully to play out, but its health impacts are certain to be considerably smaller than Chernobyl's. As there was no catastrophic failure of reactor containment, radiation exposures were considerably lower. There have been no cases of radiation sickness, though at least two workers were exposed to radiation at more than 600 millisieverts (mSv) – more than double the Japanese government's emergency limit. Another twenty-two received at least 100 mSv, the annual dose above which there is a raised risk of cancer (this extra risk is small – much smaller than smoking).

Some of these nuclear workers may well develop cancers later in life. But there is no evidence that radiation released during the accident will have wider public-health effects on residents of the surrounding area. The very highest exposures beyond the plant are estimated at less than 50 mSv per year. Parts of Europe, India and Iran experience similar natural background exposure without known adverse health effects on their inhabitants. Further afield, the risks were minimal. The French citizens who were evacuated from Tokyo will have been exposed to more radiation on their flights home than in the city.

If the health impacts of major accidents are not as frightening as anti-nuclear campaigners suggest, what of their other key objection, radioactive waste? Once again, this has been overstated. Some by-products of nuclear fission indeed remain radioactive for hundreds of thousands of years, but deploying this as an argument against atomic power brings to mind the old joke about the man asked the way to Piccadilly Circus, who replies: 'I wouldn't start from here.'

In countries that have never had a nuclear industry or an atomic weapons programme, the waste problem carries some weight as a reason not to build nuclear plants, though its significance must be balanced against the necessity of containing greenhouse emissions.

Yet in countries like Britain, France, China and America, which have had both, there is a legacy problem of radioactive waste that must be dealt with whether new nuclear plants are constructed or not. As the latest designs produce a fraction of the waste volume of older models, they would add a little to an existing problem, rather than creating an entirely new one. As deep underground disposal is generally considered the best solution, the most likely outcome is that we will just have to dig a slightly bigger hole.

The green overreaction to Fukushima also paid no heed to the unique circumstances that allowed the accident to happen and then to get out of control. The plant actually survived one of the five most powerful earthquakes on record remarkably well, but was then compromised by a tsunami much bigger than it had been designed to withstand. A higher sea wall, or an auxiliary power source raised off the ground, would likely have allowed cooling to continue and contained the crisis rapidly. Many of the activists who have used the disaster to campaign against nuclear new-build come from countries that are not seismically active or at high risk of tsunamis. Germany has a short coastline and few earthquakes. The UK has never experienced a quake of a magnitude bigger than 6.1.

Fukushima Daiichi's reactors, too, were especially vulnerable to disaster. All six were boiling water reactors (BWRs), designed in the 1960s and built in the 1970s, and the safety of their containment was not considered top-notch even then. These reactors have the distinct disadvantage that they must be actively cooled. Water must be pumped around the core even after shutdown to prevent them from overheating, which leaves them in danger in the event of a sustained loss of power. The modern reactors that are being built today have a completely different design, with passive safety systems: natural processes such as gravity and convection are used to cool them in the event of an accident so that they 'fail safe'. In an accident, they cannot help but cool themselves down, so the risk of a Fukushima-type incident is greatly reduced.

Not only did the anti-nuclear lobby ignore this important detail; many activists played up comparisons with the Chernobyl disaster, which was much more severe because of a series of design flaws unique to Soviet nuclear plants. Its graphite moderation system, coupled with water cooling, allowed pressure and heat to build up unchecked, causing a steam explosion. The reactor, unlike all those built in the West, had no substantial containment vessel, so there was nothing to stop radioactive material from escaping. The graphite in its core also caught fire, throwing a plume of radioactive debris thousands of metres into the air and distributing fallout over much of Europe. None of these factors applied in Japan – not that you would have learned this from Greenpeace or the more sensational press reports.

What Fukushima demonstrated was that running obsolete nuclear plants in the world's most seismically unstable zones, with inadequate tsunami defences, may be unwise. It has little relevance to much safer reactors in much safer parts of the globe. As George Monbiot, one of the few prominent greens with an open mind about nuclear power, wrote after the disaster: 'Using a plant built 40 years ago to argue against 21st-century power stations is like using the Hindenburg disaster to contend that modern air travel is unsafe.'

To Monbiot, Fukushima was not proof of the dangers of nuclear power, but of its safety. 'As a result of the disaster at Fukushima, I am no longer nuclear-neutral, I now support the technology,' he wrote. 'A crappy old plant with inadequate safety features was hit by a monster earthquake and a vast tsunami. The electricity supply failed, knocking out the cooling system. The reactors began to explode and melt down. The disaster exposed a familiar legacy of poor design and corner-cutting. Yet, as far as we know, no one has yet received a lethal dose of radiation.'

This is an attitude that more greens are going to have to embrace if they're really serious about containing climate change. In the UK, about 18 per cent of electricity comes from nuclear power stations, all but one of which are due to be decommissioned by 2023. Even if government targets of generating 20 per cent of electricity from renewable sources by 2020 are met, this would simply replace the low-carbon nuclear power, barely denting carbon emissions from fossil fuels. Nuclear is one of the very few reliable and mature technologies for generating electricity that is applicable almost everywhere in the world, and which does not emit significant amounts of carbon dioxide. Any dispassionate analysis of the evidence shows that if you think climate change is a problem, you should be thinking nuclear.

The green movement's position, however, is not based on evidence. Its opposition is philosophical. The environmental movement has powerful historical links with the nuclear disarmament movement, and finds it hard to cast off this ideological baggage: Greenpeace didn't get its name for nothing.

There are a few exceptions. Besides Monbiot, green converts include Mark Lynas; Stephen Tindale, a former director of Greenpeace UK; and Chris Goodall, a Green Party activist. Many scientists with impeccable environmental credentials, including Sir James Lovelock, who developed the Gaia hypothesis that sees the earth as if it were a living organism, and James Hansen, the Nasa climate scientist, are also enthusiastic. These people, though, are often seen as heretics by mainstream environmentalists.

So integral is opposing nuclear power to green politics that Lynas has described his decision to embrace it as a 'coming out' moment. 'I've been equivocating over this for many years; it's not as if it's a sudden conversion, but it's taken a long time to come out of the closet,' he said. 'For an environmentalist, it's a bit like admitting you are gay to your parents because you're kind of worried about being rejected.'

Genetically modified politics

Another part of the green package, to which all proper environmentalists are supposed to subscribe, is opposition to the genetic modification of crops. Like nuclear power, GM food is taken by the main green NGOs and political groupings to be an intrinsic evil, rather than as a neutral technology that can potentially be deployed for both good and bad ends. Interfering with genes is seen as meddling with nature – a freakish and dangerous pursuit that cannot possibly result in anything good. This is taken to be bad for the environment, and bad for human health to boot.

The argument that GM food is unsafe to eat can easily be dismissed. There is nothing in the process of genetic engineering that ought in theory to make crops any riskier than conventionally bred new varieties, and this has been borne out by experience. GM foods have been eaten by hundreds of millions of consumers in the United States for close to two decades now, without a single documented adverse consequence. A UK government review in 2003 found nothing to suggest that eating GM produce would have harmful effects, and nothing has changed since then.

Environmental questions are a little more finely balanced. It is plausible that GM crops that are resistant to herbicide, or that make their own pesticide, might have deleterious consequences for biodiversity. Herbicide-tolerant varieties, for example, could encourage farmers to apply too much herbicide, or wipe out the weeds on which birds and insects feed. The UK's farm-scale evaluations of three herbicide-tolerant crops, conducted in 2003, indeed suggested that they were a mixed blessing. While GM maize appeared to improve biodiversity, the opposite happened in fields of modified beets and oilseed rape.

The results prompted green groups to demand an immediate ban on the cultivation of any transgenic crops. But these activists both cherry-picked from the evidence and over-extrapolated its significance. As the scientists who ran them pointed out, the farm-scale evaluations showed only that particular GM varieties, grown in particular UK conditions and using a particular agronomic regime, had a negative impact on biodiversity. Yet from that, campaigners concluded that all GM crops must be environmentally damaging, and so should not be permitted.

When looked at in the round, including worldwide evidence, the environmental credentials of the first generation of GM crops look rather better. The impact of herbicide-tolerant varieties depends greatly on the attitudes of farmers, and on how they are used. If used as an excuse to spray as much as the farmer likes, there may be a damaging effect. If deployed intelligently, so applications of herbicide can be reduced, they can have environmental benefits.

A 2010 report from PG Economics, an agricultural consultancy, found the net effect was positive: GM techniques led herbicide applications to fall by 182 million kilograms between 1996 and 2006. A further benefit has been to encourage no-till agriculture: fields sown with herbicide-tolerant crops do not generally need to be ploughed, reducing carbon emissions from soil and preventing erosion. GM crops engineered to make Bt, a biological pesticide, also had a good environmental outlook. The introduction of Bt cotton has reduced applications of insecticide by 170 million kilograms.

It is possible to manage GM crop use to maximize these benefits. Varieties can be licensed for use only if a particular spraying regime is mandated. Or if herbicide-tolerant maize looks beneficial while beet does not, the former can be approved and the latter blocked. The mainstream green approach, however, has been to seize on any evidence of environmental harm from single varieties to insist on a comprehensive ban. This is rather like calling for all painkillers to be outlawed because Vioxx can have dangerous side-effects. Case-by-case regulation, so that individual crops are assessed according to their merits and risks, is not good enough for most greens. Only outright rejection of an entire application of science is seen as consistent with good stewardship of the planet.

This green intransigence is unfortunate because, just as nuclear power is probably essential to containing climate change, GM crops are likely to be pivotal to solving several other contemporary environmental challenges.

The first of these problems is land use. The world's population reached 7 billion last year and it is forecast to grow to 9 billion by 2050. If we are to stand a chance of feeding so many people sustainably, we are going to have to increase the yields we get from existing agricultural land; the alternative is to bring more and more wilderness under cultivation. GM techniques promise to be an important part of the solution, allowing farmers to get more out of their fields and to produce crops with improved nutritional qualities. Improving crop yields this way may bring regional and global environmental benefits even if biodiversity is damaged on a local level. Getting more out of existing farmland, even at the expense of farmland biodiversity, is preferable to ploughing up forest and savannah.

Genetic engineering may also be necessary if agriculture is to significantly reduce its use of other important natural resources. As most plants cannot fix the nitrogen they need from the air (the exceptions are pulses and clover, which use symbiotic bacteria to achieve this), conventional farming relies on large applications of nitrogen fertilizer. This nitrogen has damaging effects when it washes into watercourses, causing blooms of algae and depleting oxygen to create aquatic 'dead zones'. Crops that fix their own nitrogen, or which use it more efficiently, would alleviate this environmental impact. Genetic engineering is the only realistic way of achieving this. GM techniques are also likely to be important to creating new varieties that use less water and can thrive in the warmer, drier conditions we expect in many regions as a result of climate change.

GM isn't a 'silver bullet', and it won't be the solution to every agricultural problem that the world faces. But it is very likely to be among them, and we need every tool that science has to offer. It's foolish to reject it out of hand because it doesn't fit some environmentalists' idea of what is natural. The UK government's *Foresight* report into the future of food and farming, published in 2011, took an enlightened view of this after dispassionate evaluation of the science. 'New technologies (such as the genetic modification of living organisms and the use of cloned livestock and nanotechnology) should not be excluded *a priori* on ethical or moral grounds, though there is a need to respect the views of people who take a contrary view,' it concluded. It remains to be seen whether green activists will allow it to be translated into policy.

The omens aren't particularly good. When the UK began its farm-scale trials of some GM crops, to gather evidence that might answer the legitimate questions raised about the environmental impact of herbicide-tolerant varieties, the response of many mainstream green groups was to wreck them. Lord Melchett, then executive director of Greenpeace UK, was among those arrested in 1999 for ripping up a trial plot of GM maize. More recent crop trials, of potatoes modified to resist late blight, have taken place behind tight security, adding significantly to their cost. Far from being guided by the science on genetic engineering, these greens prefer science not to take place at all.

Onerous European Union regulations introduced in response to green GM protests, and the reluctance of public funders to support controversial science, also have the perverse effect of concentrating the technology in the hands of the large multinational companies to whom these campaigners most object.

'It is difficult to collect evidence of benefits or risks, given the routine destruction of GM-crop field trials by NGOs opposed to the use of the technology,' said Joyce Tait, of the University of Edinburgh, and Guy Barker, of the University of Warwick, in September 2011. 'It is difficult to develop new GM products that could be beneficial for the environment or contribute to food security when there is a lack of funding for basic research and development to produce such products. It is impossible for small companies to develop GM crops, as is generally advocated by the public, when the cost of regulatory requirements is so high that only large, multinational companies can afford it.'

Green attitudes to science are similarly selective over organic farming, an approach to agriculture that meets with the approval of environmental NGOs because it eschews pesticides and herbicides that do not occur naturally. There is good evidence that, as used in rich countries like Britain, this approach is somewhat better for the local environment than conventional farming. But it also generates lower yields, which means it would necessarily require more land. Research led by Tim Benton, of the University of York, suggests that switching all UK agriculture to organic would double the land area required for the same output. This important part of the ecological calculus is conveniently forgotten when greens campaign for wider adoption of organic techniques.

Organic lobbyists also like to argue that such food is healthier than conventionally grown produce. Evidence for this is again lacking. A large systematic review led by Alan Dangour, of the London School of Hygiene and Tropical Medicine, investigated the issue in 2009. It found no reliable indications that organic foodstuffs had any more nutritional value than their conventional counterparts. There is a good argument, indeed, that falling for organic propaganda can actually be bad for you. Science has established beyond doubt that a diet rich in fruit and vegetables is beneficial to health, yet organic produce is much more expensive than its conventional counterpart. If families on a budget choose organic fruit and veg in the misplaced belief that it is a healthier option, but buy less of it as a result, the potential for perverse consequences is clear.

The knee-jerk ideological opposition to GM crops and to nuclear power that characterizes so much of the green movement matters because it makes important technologies, with much to contribute to sustainable development and containing climate change, more difficult for governments to back. Green pressure, for example, ensured that nuclear power was excluded from the Clean Development Mechanism, the provision in the 1999 Kyoto Protocol for rich countries to offset their carbon emissions by investing in low-carbon energy in developing countries.

It also has another effect. By so transparently rejecting scientific consensus on both issues, greens invite the charge of hypocrisy when they urge politicians and the public to listen to the scientific consensus on climate change. If they are prepared to cherry-pick scientific evidence to suit their purposes on nuclear power and biotechnology, people are bound to wonder whether they are doing the same over global warming.

Science or ideology?

This green ambivalence to science has emerged because the movement's thesis on climate change is not grounded in science alone, but forms part of a much broader agenda. It is an agenda that may well begin to explain why so many members of the public, and the politicians who want their votes, are so quick to sympathize with criticisms of climate science.

For many green activists, true environmentalism is necessarily political, allied to deep suspicion of the modern technology and free-market economics that, they argue, have created many of the ecological challenges that face the world today. It is not sufficient to see climate change as a problem described by science, to which science will provide evidence-based technological solutions. It is also an opportunity to effect fundamental political and social change. It means fighting capitalism, globalization and large corporations. And it means embracing more natural lifestyles while unwinding some of the technological and industrial progress of the past two centuries.

This significant strand of green opinion is not interested in economic growth: Britain's Green Party contends that 'a society less dedicated to material growth will not only avoid ecological collapse but also make us more content'. It argues that individuals must curtail their aspirations and ambitions for the sake of the environment, and embrace austerity: we must eat less meat, turn down our thermostats, avoid flying anywhere on holiday, and even have fewer children. Jonathon Porritt, the influential environmentalist, describes couples who have more than two children as irresponsible. These messages appeal both to left-wingers who would once have become Marxist agitators and to a certain kind of conservative who romanticizes traditional ways of living. Neither Swampy nor the Prince of Wales cares much for consumerism.

There is no place here for a dispassionate, method-blind approach to mitigating and adapting to climate change by any means necessary. The solutions must be politically correct. They must involve fundamental behaviour change that reduces consumption, particularly in rich countries, and they must not involve technology that is perceived as 'unnatural'. GM crops and nuclear power are rejected as 'techno-fixes' that involve human manipulation of the natural world and serve the interests of big corporations. Emerging technologies such as synthetic biology and nanotechnology, and future developments such as nuclear fusion and geo-engineering, are similarly portrayed not as opportunities but as threats.

These technical solutions are opposed, feared even, because they have the potential to allow humanity to continue with something approaching business as usual, allowing sustainable growth and economic development without requiring significant changes to consumer and corporate behaviour, or to the capitalist system. While they might make important contributions to containing climate change and other environmental problems, they do nothing to advance the political element of the green agenda, and may even work against it. 'The reason why nuclear power is so heavily opposed by the greens is not because it can't help to solve climate change, but because it can,' says Lynas.

> *'The reason why nuclear power is so heavily opposed by the greens is not because it can't help to solve climate change, but because it can' –*
> *Mark Lynas*

This alignment of climate change with a wider political agenda is dangerous. As well as sowing unnecessary controversy around promising technologies, it makes it considerably harder to convince both ordinary people and politicians of the importance of tackling climate change. The position that global warming is a planetary emergency that requires all the weapons in our arsenal – including investment in nuclear, GM and other new technologies as well as costly reductions in carbon emissions and personal behaviour change – is probably just about a saleable political proposition. The idea that we must eschew growth and modern technology, abandon driving and flying, and live off what we can produce locally, is not.

By using the science of climate change to advance a social and economic manifesto, the mainstream green movement has also needlessly made enemies of much of the political right. The science of global warming ought not to be a political issue: it is at root a matter of physics and chemistry, which shouldn't look different to conservatives and liberals. Yet a large majority of those who question the evidence for climate change label their political views as right wing.

This is no accident. The politicization of the science has invited those who disagree with the political solutions advanced by the green movement to go after the science as well. Frank Luntz wasn't asked to advise the Republicans on spreading doubt about global warming because there is something inherently left wing about carbon dioxide's effects on the atmosphere. Greens have so successfully linked the science of climate change to a particular set of policy solutions, with which conservatives are uncomfortable, that fighting the science became a way to undermine that political narrative. 'The right-wing climate contrarians and the greens actually agree that climate change means we have to dismantle industrial civilization,' says Lynas. 'The greens want this, the right wing doesn't. It leads them both to manipulate the science.'

Daniel Sarewitz, Professor of Science and Society at Arizona State University, has argued persuasively that green tactics effectively invited right-wingers to refuse to accept the scientific consensus. 'Think about it,' he wrote in an article for *Slate*. 'The results of climate science, delivered by scientists who are overwhelmingly Democratic, are used over a period of decades to advance a political agenda that happens to align precisely with the ideological preferences of Democrats. Coincidence – or causation? Now this would be a good case for *Mythbusters*.'

Conservatives, particularly in the US, instinctively distrust the type of institutions that environmentalists venerate and the policies they promote, Sarewitz notes. The Kyoto Protocol was a top-down initiative, run under the auspices of the United Nations, which required the citizens of developed countries to make economic sacrifices while emerging rivals such as China and India made none.

It could almost have been designed to make a certain kind of Republican neuralgic.

Steve Rayner, James Martin Professor of Science and Civilization at Oxford University's Saïd Business School, makes the further point that for the past decade greens have demonized those who oppose the Kyoto Protocol as enemies of science. They have thus started to behave as enemies, attacking the science describing the problem because they dislike the politics of the advertised solution.

'It became that if you buy the diagnosis, you buy the prescription,' says Rayner. 'So people who didn't like the prescription started to reject the diagnosis. It's dangerous for science to become a surrogate for political debate.'

Sir David King, who as chief scientist to the UK government described climate change as a greater global threat than international terrorism, says this analysis is 'a cogent description of what has happened. The NGOs and even scientific bodies like the IPCC have become far too closely tied to the Kyoto process. As a result, we've not been able to deal with the politics of an industrializing China and India and the US rejection.'

It ought to be perfectly possible to accept the science of climate change and to reject the particular solutions advanced by Kyoto. But green activism has made this difficult. The result has been to create a powerful constituency that delightedly seizes on any indication, however poorly founded, that the science of climate change might be less than robust. If the problem can be discredited, then so can the solutions.

Green geeks

There are lessons here for the geek movement. As we seek to secure a greater role for evidence and the scientific method in the political process, we have to stick to the evidence we care about so much, and avoid blurring the lines between science and politics. As Evan Harris says: 'We are held back by the rationality and circumspection with which we speak, handicaps that do not encumber our opponents.' Yes, we need a compelling narrative, we need to tell our story in an accessible and arresting manner. But if we're going to fight for and with science, we have to be true to it.

There is also an important role for geek activism in promoting a new kind of environmental politics, grounded in science rather than ideology. We must be robust campaigners for what the science really shows. It isn't enough for climate scientists to refute their critics in the pages of peer-reviewed journals, as Phil Jones would prefer. All of us who understand what the peer-reviewed literature shows need to explain it cogently at every opportunity and to challenge those who question it in public forums.

Climate sceptics often sow doubt about the science of human-induced global warming by attacking single pieces of evidence that have been advanced in its support, such as the errors identified in the IPCC report. We need to communicate more effectively that the scientific consensus is not threatened by individual mistakes or controversies, but is a broad conclusion drawn from multiple branches of research. The theory is consistent with the evidence from a multitude of different fields. Ice-core research in Antarctica, patterns of species distribution among animals and plants, measurements of temperature and Arctic ice extent from both satellites and the ground, and results from many other disciplines all point the same way. Thousands of scientists could just possibly be mistaken – but from so many different fields?

There is an analogy worth making with another branch of science that causes unwarranted controversy, if to a lesser extent in Europe than in the US. The fact of evolution is also based on evidence not from one discipline, but from twenty or more. If the principle is wrong, then so is much of zoology, botany, palaeontology, genetics, molecular biology, geology, anthropology and medicine, to name but a few fields through which it traces a common thread. Paul Nurse explained this well in his *Horizon* documentary: science, he said, works by examining how evidence fits together as a whole. It's an argument that could be better deployed to explain why the consensus is so powerful.

We can also take the fight to the deniers and sceptics by exposing how many of them have both attacked other findings of science that have political implications, and moved the goalposts as the science of climate change has become more and more settled. Fred Singer, a retired American physicist, has long been at the forefront of the dwindling group of scientists who dispute anthropogenic climate change. In their 2010 book *Merchants of Doubt*, Naomi Oreskes and Erik Conway revealed how Singer and several colleagues have fought to undermine other scientific findings that were damaging to certain business interests, such as the link between smoking and lung cancer and the damage to the ozone layer caused by chlorofluorocarbons.

Singer and the contrarians also have a history of shifting their positions, as anomalies they highlight in the climate data are resolved. For many years, they pointed to discrepancies between temperatures in different parts of the atmosphere. When these were reconciled, they changed tack, suggesting that the world might be warming, but solar activity or cosmic rays might be responsible. At present, a favoured line of argument is that the warming has stopped. These are the tactics of denialists who start from a conclusion, then go in search of supporting data.

Oreskes and Conway argued that mainstream scientists have been too slow to make these points, or have done so ineffectively. 'Scientists are finely honed specialists trained to create new knowledge, but they have little training in how to communicate to broad audiences, even less in how to defend scientific work against determined and well-financed contrarians. They often have little talent or taste for it either. Until recently, most scientists have not been particularly anxious to take the time to communicate broadly. They consider their "real" work to be the production of knowledge, not its dissemination, and they often view these two activities as mutually exclusive. Some even sneer at colleagues who communicate to broader audiences, dismissing them as "popularisers".'

This has to change if the critics are effectively to be countered. We also need to show our support for scientists who find themselves targeted by sceptics. As Phil Jones's experience shows, it can be highly distressing to see one's work denigrated unfairly by vituperative bloggers. We need to tell these scientists that what they do is appreciated.

The climate sceptics, though, aren't the only voices that need to be countered by the geeks. We also need to redress the damaging messages that the greens have promoted, that environmentalism can take or leave science to suit its purposes, and cares as much about challenging capitalism and limiting the spread of certain technologies as it does about containing climate change and other ecological threats.

We could begin by targeting the green movement, to try to convince the influential NGOs like Friends of the Earth and Greenpeace to adopt a different strategy. Most geeks count themselves environmentalists, and thousands have joined these organizations in the belief that they can be trusted to advance environmental protection. Many are deeply frustrated by the positions they take on nuclear energy, GM crops and other scientific issues, yet continue to pay subscriptions because they think the importance of climate change demands strong campaigning. Those geeks who are Greenpeace members need to share their views with the leadership. Changing the NGOs' direction will be a tall order, but it is never going to happen if they do not believe there is demand from their supporters. If the message goes unheeded, resign your membership, and take care to explain why.

This sort of criticism can sometimes pay dividends. Before the European elections in 2009, the science bloggers Martin Robbins and Frank Swain examined the science policies included in the various party manifestos and were shocked by some of the positions taken by the Green Party. Alongside predictable opposition to nuclear power and GM crops were plans to ban stem cell and much animal research, and strong support for alternative medicine. When they published their criticisms, Robbins was invited to attend the Green Party conference, where he worked with a group of activists to draw up more sensible policies on these three aspects of science. His critical approach helped party members who despaired of some of their leadership's policies to change them.

If the NGOs won't change, then there's a good case for founding a new pressure group that takes a strongly evidence-based approach to environmental issues, and which decouples them from the austere, anti-growth, back-to-nature politics with which it has become indelibly associated. As Lynas says, environmentalism has to lose this baggage. 'I want an environmental movement that is happy with capitalism, which goes out there and says yes rather than no, and is rigorous about the way it treats science. The green movement needs a clause-four moment – the Labour party had to go through that.'

Even without a formal organization through which to work, this is a goal for which individual geeks should be working on our own initiative. All the political and public lobbying for which this book has been arguing – the letter-writing and emailing, the meeting MPs, the joining political parties – can be used in support of this important environmental case. Politicians need to know that people who care about climate change and the environment are not limited to those who subscribe to other aspects of the green agenda. We have to make them understand that following the evidence on issues like nuclear power and GM crops can win as well as lose support, and that pandering to the louder greens can carry a cost.

Our society's chances of mitigating and adapting to climate change, indeed, would be considerably improved by all the themes *The Geek Manifesto* has articulated. We need politicians who understand how science works, and who use evidence responsibly to develop policy. We need a media that looks beyond phoney balance, and that covers evidence rather than hearsay. We need proper acknowledgement of how green technology can contribute to economic growth, and an education system that introduces children properly to the scientific method. And we need geeks to use our newfound political confidence to step up the pressure.

Science has described what may well be the greatest political challenge of the coming decades. It is going to be essential to meeting it.

GEEKS OF THE
WORLD UNITE

IN THE EARLY STAGES OF THE FRENCH REVOLUTION, IN 1789, KING Louis XVI was forced to recognize a new National Assembly. The landed supporters of the Ancien Régime chose to sit to the right of the president's chair; their opponents gathered to the left.

Thus began a taxonomy of modern politics that has endured across many different cultures for more than two centuries. The right wing defends capital and free enterprise, supports traditional institutions and dislikes challenges to social norms. The left campaigns for egalitarian economics while embracing social change.

This binary notion of politics is increasingly regarded as out of date. We all know it's perfectly possible to combine a strong belief in free-market economics with a permissive stance on drugs or homosexuality. Many religious conservatives who think premarital sex and abortion are sinful believe passionately in reducing inequality, while plenty of socialists are tough on crime.

It's now widely accepted that opinions about social and economic issues do not necessarily go hand in hand. A fascinating website, the Political Compass, will even plot your views on two axes – one measuring your attachment to free-market economics, and the other your social libertarianism. I come out a little right-of-centre on economics, and as a fairly strong social liberal.

Even this refinement, though, is no longer enough. Politics has a third axis too. It measures rationalism, skepticism and scientific thinking – the willingness to base opinions on evidence and to keep them under review as better evidence comes along.

At one end of it stand the geeks, who otherwise belong to all sorts of political tribes. We have too few politicians of any persuasion for company. It shows in the way they run our countries. As we've seen throughout this book, indifference to science and outright science abuse are found across the political spectrum.

A Labour government sidelined the scientific evidence on drugs, and the actions of a Labour prime minister encouraged a vaccine scare. The Conservative–Liberal Democrat coalition that succeeded it cut the science budget and introduced a damaging immigration cap. Tom Harkin, a senior Democrat, is the US Senate's leading champion of alternative medicine; James Inhofe, a senior Republican, is its leading champion of climate denial.

Ministers of all parties, and public servants of none, show a troubling reluctance to use the methods of science to investigate how best to teach children, prevent crime, fund healthcare or protect the environment. Many of them would prefer it if the policies they implement were never evaluated at all.

They do it because they know no better. And they do it because we let them. If geeks want change, we can't continue to sit back and be ignored. We must distil our enthusiasm and outrage into a political force that punches its weight.

That's starting to happen. When Jenny Rohn wrote the blogpost that triggered the Science is Vital protests, she titled it: 'In which the great slumbering scientific beast awakens'. She was right: something large and powerful is indeed beginning to stir. In that campaign, in the backlash against the sacking of David Nutt, in the quacklash that helped Simon Singh to see off the chiropractors, geeks have shown what a little political engagement can achieve.

It's time to take it further. Let's build a movement.

This geek movement shouldn't be built around support for any particular policy, though there will be many issues, such as homeopathy and climate change, about which most of us agree. Its central concern must be how those policies are put together.

- We want a political culture that appreciates the power of science as a problem-solving tool and that seeks to exploit its methods of inquiry to resolve the great questions of the day. We want leaders who don't want just to implement their ideas, but to test them, and who are comfortable with changing their minds.

- We want politicians to listen to scientific evidence and advice with respect, and to consider it properly before they act. While we'd like them to heed it, we accept that won't always happen. But we expect them to be honest about their reasons when they go their own way.

- We want governments to create an environment in which science can thrive, to create life-enhancing technologies and to drive economic growth. We want long-term investment in curiosity, which isn't interrupted when times are tight. The evidence suggests it will pay off.

- We want a media that critically evaluates scientific claims according to the evidence behind them, and that avoids damaging hype about breakthroughs and scares. It should give science the same respect it gives to business, sport or the arts.

- And we want science and critical thinking to become central to the national conversation. We want as many people as possible to appreciate not only what science achieves, but how it achieves it. We know that that has to start at school.

If we're to achieve these goals – and they won't come easily – we have to raise the political profile of science. Every geek has a part to play.

- We must challenge politicians whenever science is twisted, misused or simply passed over, and we need to do it in numbers. There are millions of us out there. We have the online tools to assemble. Let's make ourselves heard. We must use our voices, and use our votes. Science abuse must carry a cost. Then we can become a constituency to be courted and appeased, not ignored.

- We must lobby our elected representatives, taking time to explain how and why evidence matters. Geeks are constituents too – we have a right to expect politicians to listen to us, even if they won't always agree. When governments hold public consultations, we must participate. When they don't, we must share our views anyway. We can convince MPs and congressmen to take up our grievances, and help them to help us hold governments to account.

- We must get more involved. Too few geeks join political parties and play a part in selecting the candidates between whom voters choose. Still fewer of us stand for office. Those of us who support a party should sign up so we can influence its positions from the inside.

- We must use the media and work to improve it. Let's support both scientists and journalists who communicate science well, and complain constructively about those who don't. With blogging and social media, we can also set the record straight ourselves.

- We must use every means available to us to promote a better appreciation of scientific thinking. We can volunteer in schools. We can put pressure on green groups we belong to, companies we buy from and charities we support. We can ask businesses for evidence to support their claims and complain to regulators when they refuse. And we must discuss our views freely with our families, colleagues and friends, instead of keeping quiet for fear of causing offence.

- In all this, we must stay true to the values we want to promote. As Evan Harris says, we mustn't be tempted to throw off the constraints of rationality and circumspection: it's these qualities that give science its power. Compromise here, and we risk becoming just another special interest.

In 1950, Robert H. Jackson, a US Supreme Court Justice, dissented in part from a judgment that union leaders could be forced to swear an anti-communist oath. 'It is not the function of our Government to keep the citizen from falling into error,' he said. 'It is the function of the citizen to keep the Government from falling into error.'

That is a function that geeks must embrace with alacrity. We understand that nothing can prevent error, but also that the methods of science build the sturdiest bulwarks against it that we've got. It's up to us to show the world why science matters.

REFERENCES

Epigraphs

Carl Sagan: 'We've arranged a global civilization . . .': The *Demon-Haunted World*, p. 26.

Bertrand Russell: 'A habit of basing convictions upon evidence . . .': quoted in G. Simmons, *Calculus Gems* (New York, 1992).

1 The Geeks Are Coming

'You might think that modern chiropractors restrict themselves . . .': 'Beware the Spinal Trap', Simon Singh, Guardian, 19/4/2008. http://www.guardian.co.uk/commentisfree/2008/apr/19/controversies inscience-health

' "We're rationalists," as Singh puts it': interview with author, 29/12/2010. Facebook group http://www.facebook.com/group.php?gid=33457048634

'In May 2009, Mr Justice Eady ruled that, by calling their claims "bogus" . . .': David Allen Green, 'Simon Singh's Bogus Journey', *Lawyer*, 23/2/2010. http://www.thelawyer.com/simon-singh's-bogus-journey/1003557.article

'the Penderel's Oak pub in Holborn, central London, thronged with geeks . . .': see 'Now charlatans will know to beware the geeks', Nick Cohen, *Observer*, 18/4/2010. http://www.guardian.co.uk/commentisfree/2010/apr/18/nickcohen-simon-singh-libel

' "The reaction was extraordinary," Singh says': interview with author, 29/12/2010.

'No-one would have thought badly of Simon . . .': David Allen Green, Jack of Kent blog, 18/4/2010. http://jackofkent.blogspot.com/2010/04/bca-v-singh-did-skeptics-really-make.html

'When the BCA released what it called a "plethora of evidence" . . .': http://www.chiropractic-uk.co.uk/gfx/uploads/textbox/Singh/BCA%20Statement%20170609.pdf

'Then there was what Green dubbed the "quacklash" . . .': David Allen Green, Jack of Kent Blog, 13/6/2009. http://jackofkent.blogspot.com/2009/06/quacklash-causes-and-effects.html

' "I don't think there could be a better use of £75 worth of stamps," wrote Perry': '500 Chiropractors reported to Trading Standards and GCC', Simon Perry, Adventures in Nonsense blog, 13/6/2009. http://adventuresinnonsense.blogspot.com/2009/06/500-chiropractors-reported-to-trading.html

'Complaints about more than 500 individual practitioners in just 24 hours': ' "Witch hunt" forces chiropractors to take down their websites', Chris French, *Guardian*, 20/6/2009. http://www.guardian.co.uk/science/2009/jun/19/chiropractic-bca-mca-singh

'Lewis got hold of an email from the McTimoney Chiropractic Association urging its members to take down their websites . . .': 'McTimoney Chiropractors told to take down their websites', Andy Lewis, 'The Quackometer' blog, 10/6/2009. http://www.quackometer.net/blog/2009/06/chiropractors-told-to-take-down-their.html

'In April 2010, the Court of Appeal overturned Mr Justice Eady's ruling in a withering judgment . . .': Full judgment: http://www.bailii.org/ew/cases/EWCA/Civ/2010/350.html

'One in four British chiropractors was under investigation by regulators at the time': 'Furious backlash from Simon Singh libel case puts chiropractors on ropes', Martin Robbins, *Guardian* blogs, 1/3/2010. http://www.guardian.co.uk/science/2010/mar/01/simon-singh-libel-case-chiropractors

' "Scientific controversies must be settled by the methods of science rather than by the methods of litigation,' the judgement noted': Court of Appeal, BCA vs Singh, para 34 http://www.bailii.org/ew/cases/EWCA/Civ/2010/350.html

'principled scepticism': for a discussion of 'skeptics with a k', see http://www.ukskeptics.com/article.php?article=skeptic_or_sceptic.php&dir=articles

'Sense About Science began a "Keep Libel Laws Out of Science" petition . . .': http://www.libelreform.org/

'which still left a vindicated Singh £60,000 out of pocket because of legal costs he could not recover . . .': *The Times*, 16/4/2010: 'Science writer Simon Singh wins bitter libel battle', Mark Henderson, Science Editor. http://business.timesonline.co.uk/tol/business/law/article7098157.ece

'In March 2011, the Conservative–Liberal Democrat coalition introduced a draft defamation bill': http://www.justice.gov.uk/consultations/docs/draft-defamation-bill-consultation.pdf

' "Science is more than a body of knowledge," said Carl Sagan': *The Demon-Haunted World*, p. 25.

'In the metaphor of the skeptical writer Michael Shermer, it is not a noun but a verb': 'If you could teach the world just one thing', *Spiked Online*. http://www.spiked-online.com/articles/0000000CAAA0.htm

' "Science is a way of trying not to fool yourself," explained Richard Feynman': from lecture 'What Is and What Should be the Role of Scientific Culture in Modern Society', given at the Galileo Symposium in Italy (1964).

'as John Holdren, President Obama's chief science adviser, explains: "more and more of the public policy issues that are before us . . . have science and technology content" ': John Holdren, press conference at American Association for the Advancement of Science conference, Washington, DC, 18/2/2011.

'When Sir David King was the chief scientific adviser to Tony Blair's government . . .': interview with author, 10/3/2011.

'famously identified by C. P. Snow in his "Two Cultures" lecture of 1959 . . .': Snow's 1959 Rede Lecture is available online at http://books.google.com/books/about /The_two_cultures.html?id= OyHm4sc6IPoC

'There are fifty-five US senators with law degrees – more than half the Senate's membership. None has a PhD in the natural sciences, and only one, Chris Coons of Delaware, has an undergraduate science degree': *Membership of the 112th Congress: A Profile*, Jennifer E. Manning, Congressional Research Service. http://www.fas.org/ sgp/crs/misc/R41647.pdf

'The UK's 650 MPs include just three with science PhDs, only one of whom, Julian Huppert of Cambridge, has worked in research': 'Parliament will lose much scientific expertise after election', *The Times*, 27/4/2011. http://www.timesonline.co.uk /tol/news/politics/article7108998.ece (subscription required, as for all *Times* and *Sunday Times* links).

'*His Wonders of the Solar System* and *Wonders of the Universe* programmes have topped BBC2's ratings': Broadcasters' Audience Research Board, week ending 6 March 2011. http://www.barb.co.uk/report/weeklyTopProgrammesOverview

' "I liked the Class 51 that Greater Manchester used to have," he told Hugo Rifkind of *The Times*': 'Falling for Gravity', Hugo Rifkind, *The Times* (*Eureka*), 3/3/2011. http://www.thetimes.co.uk/tto/science/eureka/article 2930397.ece

'The targets of his "Bad Science" column . . .': Ben Goldacre's 'Bad Science' columns are archived at www.badscience.net

'His *Bad Science* book was a bestseller, shifting more than 300,000 copies in the UK alone . . .': figures from Nielsen BookScan.

'Dara O'Briain is a physics graduate who peppers his routine with jokes about homeopathy': see for example: http://www.youtube.com/watch?v= YMvMb90hem8

'Tim Minchin, the comic pianist, performs a nine-minute beat poem called "Storm" . . .': a version is available at: http://www.youtube.com/watch?v=UB_ htqDCP-s

' "People were waiting for something like this," he says': Robin Ince, interview with author, 24/1/11.

'The Cheltenham Science Festival, first staged in 2002, sold more than 30,000 tickets in 2011 . . .': figures provided to author by Sharon Bishop, executive producer.

'New York's World Science Festival, founded in 2008, attracts hundreds of thousands of people': *New York Times*, 03/06/2008. http://www.nytimes.com/2008/06/03/science/ 03fest.html

'In 2009, *The Times* launched *Eureka*, a successful monthly science magazine . . .': *Eureka* is published on the first Thursday of the month, and is available online at www.thetimes.co.uk/eureka

'The *Guardian* has established a science blogging network': this can be read at http://www.guardian.co.uk/science-blogs

'Dozens of British towns and cities now have branches of Skeptics in the Pub (SiTP) . . .': you can find your local branch at http://skeptic.org.uk/events/ skeptics-in-the-pub

' "There is an ever expanding army of geeks and the wonderful thing about it is its somewhat anarchic nature," says David Colquhoun': email to author, 17/1/2011.

' "We are entering the age of the geek," Cox told the *Sunday Times*': 'Brian Cox: the new Mr Universe', Eleanor Mills, *Sunday Times*, 27/2/2011. http://www.thesundaytimes. co.uk/sto/Magazine/Interviews/article567104.ece (subscription required)

'The Campaign for Science and Engineering estimates that more than three million people in Britain have some sort of science background: a relevant degree . . .': *CaSE News*, 63, April 2010. http://www.sciencecampaign.org.uk/ members/casenews/CaSENews63.pdf

'almost as many voters as all the ethnic minorities put together': Ethnic minorities account for about 8 per cent of the UK population: Ben Smith, House of Commons Library, 18/11/2008, 'Ethnic Minorities in Politics, Government and Public Life'. www.parliament.uk/briefingpapers/SN01156.pdf

'In the US, the National Science Foundation counts at least 5.5 million working scientists and engineers . . .': NSF Science and Engineering Indicators, 2010. http://www.nsf.gov/statistics/seind10/c3/c3h.htm

'Before the 2010 general election, David Cameron, Gordon Brown and Nick Clegg did everything they could to engage with Mumsnet . . .': I am grateful to Stephen Curry, Professor of Structural Biology at Imperial College London, for the idea. Incidentally, Mumsnet has been an important player in the libel reform campaign.

2 Geeking the Vote

'Tredinnick got to his feet in the House of Commons': Hansard, 14/10/09. http://services. parliament.uk/hansard/Commons/ByDate/20091014/mainchamberdebates/ part009.html

'Back in 2001, Tredinnick told the Commons': Hansard, 24/01/01 http://www. publications.parliament.uk/pa/cm200001/cmhansrd/vo010124/halltext/ 10124h02.htm

'he charged the taxpayer £755.33 for astrology software': 'David Tredinnick defends expenses claim for astrology software', *Hinckley Times*, 26/06/2009. http:// www.hinckleytimes.net/news-in-hinckley/local-news/hinckley-news/2009/06/26/ david-tredinnick-defends-expenses-claim-for-astrology-software-105367-23988439/

'[Treddinick] recently asked the Health Secretary to congratulate homeopathic chemists on their contribution to containing swine flu': Hansard, 10/01/2011; reported in http://www.theyworkforyou.com/debates/?id=2011-01-10b.23.0

Tredinnick urges health minister to be 'very open to the idea of energy transfers': Hansard, 14/10/09. http://services.parliament.uk/hansard/Commons/ByDate/20091014/ mainchamberdebates/part009.html

'[Nadine Dorries] likes to promote an urban myth about a twenty-one-week foetus grasping a surgeon's finger': Ben Goldacre, Nadine Dorries and the Hand of Hope, 19/3/2008. http://www.badscience.net/2008/03/nadine-dorries-and-the-hand-of-hope

'[Peter Hain] convinced himself that homeopathy cured his son's eczema, and promoted alternative medicine from a position of power': speech to The Prince of Wales Foundation for Integrated Health, 12/10/2005; reported in David Colquhoun, *DC's Improbable Science*, 'Peter Hain and GetwellUK: pseudoscience and privatisation in Northern Ireland', 9/2/2007. http://www.dcscience.net/?p=33

'[Anne Milton] cites her grandmother's experience as a homeopathic nurse in support of NHS funding of alternative medicine': http://skeptical-voter.org/wiki/index. php?title= Anne_Milton#Homeopathy

'[Dana Rohrabacher] took issue with conventional explanations for sharp global warming in prehistoric times': reported in http://thinkprogress.org/2007/02/10/dino-flatulence/

'[Tom Coburn] asserted that "condoms do not prevent most STDs" ': reported in Chris Mooney, *The Republican War on Science*, p. 214.

'[Tom Harkin] convinced that his allergies were cured by a supplement known as bee pollen': 'Cures or "Quackery"? How Senator Harkin shaped federal research on alternative medicine', Stephen Budiansky, 9/7/95. http://www.usnews.com/usnews/ culture/articles/950717/archive_032434.htm

'the US National Center for Complementary and Alternative Medicine, which wastes about $130m a year on studies of what you might call bogus therapies': full budget figures available at http://nccam.nih.gov/about/budget/congressional/

'[Gary Goodyear] refused to say whether he believed in evolution': 'Science minister won't confirm belief in evolution', Anne McIlroy, *Toronto Globe & Mail*, 17/3/2009. http://www.theglobeandmail.com/news/national/article320476.ece. Thanks to Baba Brinkman for bringing to my attention.

'In his famous "Two Cultures" lecture of 1959, C. P. Snow posed a challenge': Snow's 1959 Rede Lecture is available online at http://books.google.com/ books/about/The_two_cultures.html?id=OyHm4sc6IPoC

'Some 158 have a background in business and 90 were political advisers or organiz-ers': 'Social background of MPs', House of Commons Library, Feargal McGuinness, 14/12/2010. www.parliament.uk/briefing-papers/SN01528.pdf

'An analysis by *The Times* after the 2010 general election': 'Election 2010: a terrible night for science' http://www.thetimes.co.uk/tto/science/eureka-daily/?blogId= Blog3dfc20db-8d88-49bd-9347-1957bc781c72Post3c6f9697-96ae-433b-a579-9dc67ab84c11

'The situation is no better in the US': 'Membership of the 112th Congress: A Profile', Jennifer E. Manning, Congressional Research Service http://www.fas.org/ sgp/crs/misc/R41647.pdf

'Adam Afriyie, then the Tory Shadow Science Minister, pledged to hold compulsory science seminars': 'Conservative MPs will be forced to keep up with science', *The Times*, 17/11/2008. http://www.thetimes.co.uk /tto/news/uk /article 1967130.ece

'When the Parliamentary Office for Science and Technology held such a session, open to all parties, it had become optional and barely a dozen MPs turned up': email from Chandrika Nath, Parliamentary Office of Science and Technology, 12/10/2011.

'A recent participant in the Royal Society's pairing scheme, which links scientists and politicians, repeatedly forgot his pair's name and described him as "the work experience" ': conversation with someone else involved in the scheme, who requested that no parties be named.

'When three UK scientists won Nobel Prizes in 2010, he omitted to congratulate them': Roger Highfield, *New Scientist* S-Word blog, 6/10/2010, 'David Cameron: Science isn't even on his radar' http://www.newscientist.com/blogs/thesword/2010/10/david-cameron-science-isnt-eve.html

'[Obama's] budget compromise agreed in summer 2011 seemed certain to cut the funds available for research': 'Senate Panel Trims NIH Budget By $190 Million', Jocelyn Kaiser, *Science Insider*, on 20/9/2011. http://news.sciencemag.org/scienceinsider/2011/09/senate-panel-trims-nih-budget-by.html; 'Senate Panel Cuts NSF Budget by $162 Million', Jeffrey Mervis, *Science Insider*, 14/9/2011. http://news.sciencemag.org/scienceinsider/2011/09/senate-panel-cuts-nsf-budget-by.html

'In 2008, indeed, both [Obama] and his main opponent in the Democratic primaries, Hillary Clinton, refused an invitation to debate science at the Franklin Institute in Philadelphia': ScienceDebate2008; Chris Mooney and Sheril Kirshenbaum, *Unscientific America*, pp. 53–55.

'John McCain, the Republican front-runner, didn't even bother to reply': both Obama and McCain did eventually reply to fourteen written questions posed by the ScienceDebate 2008 campaign. Their responses can be found at http//:www.sciencedebate.org/debate08.html

'The demands of lab and lecture theatre mean that students reading medicine, engineering or physics typically have twice as many scheduled teaching hours as their peers studying history or languages': 'The academic experience of students in English universities', Bahram Bekhradnia, Carolyn Whitnall and Tom Sastry, Higher Education Policy Institute, October 2006. http://image.guardian.co.uk/sys-files/Education/documents/2006/10/30/ Hepireport.pdf

'the University of Bristol tells prospective arts students': 'Studying arts at Bristol' http://www.bris.ac.uk/arts/current/under/study.html

'what Carl Sagan called "the hard but just rule"': *The Demon-Haunted World*, p. 32.

'Richard Dawkins is fond of recounting an anecdote': Dawkins, *The Root of All Evil*, Part 1, broadcast January 2006, Channel 4. http://video.google.com/videoplay?docid=9002284641446868316

'For evidence, look no further than Senator John Kerry': for a fuller account, see Kathryn Schulz, *Being Wrong* (Portobello, 2010), p. 176.

'[John Kerry] will be forever haunted by his remark about funding for the Iraq war: "I actually did vote for the $87 billion before I voted against it" ': quoted in 'John Kerry's Top Ten Flip-Flops', Joel Roberts, CBS News, 2/11/2009. http://www.cbsnews.com/stories/2004/09/29/politics/main646435.shtml

'Tim Harford, the *Financial Times* columnist and author of *The Undercover Economist*, explains the drawbacks eloquently': interview with author, 25/1/2011.

'[Bill Foster says] "It's a nasty job, politics"': interview with author, 18/2/2011.

'Richard Feynman famously described': 'What is and What Should be the Role of Scientific Culture in Modern Society?', Galileo Symposium, 1964.

'"Remembering to attend to counter-evidence isn't difficult, it is simply a habit of mind," wrote Kathryn Schulz in Being Wrong': Schulz, Being Wrong, p. 132. Pp. 128–32 give an excellent account of cognitive dissonance and confirmation bias.

'"They want the authority of the white coat," says Alan Leshner': interview with author, 18/2/2011.

'Michele Bachmann regards climate change as a hoax based on "manufactured science"': http://www.youtube.com/watch?v=JUVUX3F72Xw

'Rick Perry holds similar views': Fed Up: Our Fight to Save America from Washington!, Rick Perry (Little, Brown, 2011).

'Both candidates also dismiss the fact of evolution': Perry: 'Rick Perry: Evolution is "theory" with "gaps"', Catalina Camia, U.S. Today, 28/8/2011. http://content.usatoday.com/communities/onpolitics/post/2011/08/rick-perry-evolution-presidential-race-/1

'Bachmann: Schools should teach intelligent design', CNN, 17/6/2011. http://politicalticker.blogs.cnn.com/2011/06/17/bachmann-schools-should-teach-intelligent-design

'Bachmann has even invented fictional Nobel laureates': in this interview: http://www.youtube.com/watch?feature=player_embedded&v=Damah0KH-Co

'Mitt Romney has backtracked from previous statements that climate change is happening and humans are responsible': 'Romney tweaks his climate-change stance', Shira Schoenberg, Boston Globe, 29/8/2011. http://www.boston.com/Boston/politicalintelligence/2011/08/romney-tweaks-his-climate-change-stance/oUWhiMkikNwcSqN062CHZL/index.html

'To be clear,' Huntsman tweeted, 'I believe in evolution and trust scientists on global warming. Call me crazy': http://twitter.com/#!/JonHuntsman/status/ 104250677051654144

'Three past or present MPs are even trustees of a campaign group': Joe Benton (Labour, Bootle), Ian Gibson (Labour, formerly Norwich North) and Caroline Lucas (Green, Brighton Pavilion) are trustees of the EM Radiation Research Trust. http://www.radiationresearch.org/index.php?option=com_content&view=article&id=7&Itemid=12

'[Bill Foster] remembers a Democratic colleague who argued that "we should really put more windmills into our energy policy because they poll well"': interview with author, 18/2/2011.

'Kevin Eggan's lab at Harvard introduced a sticker system': 'The "untouchables" of US science', Ed Pilkington, Guardian, 29/12/2006. http://www.guardian.co.uk/science/2006/dec/29/genetics.research

'Academic institutions, however, were now being given strict quotas on the number of visas they could issue, which severely limited their capacity to recruit': 'Migration cap costs Britain the services of 233 top brains', Mark Henderson, Abi Millar, Rebecca Hill, The Times, 9/11/2010. http://www.thetimes.co.uk/tto/science/article2801532.ece

'The Institute of Cancer Research in London, which employs scientists from fifty-five countries and is Britain's top-rated academic institution, was told its visa allocation was being cut from thirty to four': 'Immigration cap threatens cancer research in Britain', Mark Henderson, *The Times*, 3/11/2010. http://www.thetimes.co.uk /tto/science/medicine/article2791819.ece

'Eight Nobel laureates wrote to *The Times* to protest that the measure was damaging British research': 'UK must not isolate itself from research world', *The Times*, 7/10/2010. http://www.thetimes.co.uk/tto/opinion/letters/ article2755952.ece

'A 2008 study found that the Act's requirements cost one university £51,000 a year': 'Delivering innovative cancer diagnostics and treatments to patients', *Newton's Apple*, July 2008. http://www.newtons-apple.org.uk/Delivering_Innovative_Cancer_Diagnostics_ and_Treatments_to_Patients.pdf

'Peter Furness, vice-president of the Royal College of Pathologists, told the *Guardian*': quoted in 'Rules on use of human tissue stifle research, say scientists', James Randerson, *Guardian*, 18/7/2008. http://www.guardian.co.uk/science /2008/jul/18/medicalresearch

'Sir Michael Rawlins, who led an Academy of Medical Sciences review of regulation published in 2011, described it as a disaster': 'Complex and excessive rules hampering health research, report warns', Mark Henderson, *The Times*, 1/1/2011. Rawlins: http://www.thetimes.co.uk/tto/health/news/article2869923.ece

'The Academy's report revealed how researchers must often seek similar permission from a dozen or so different bodies before embarking on research involving patients as subjects, when one or two would suffice': 'A new pathway for the regulation and governance of health research', *Academy of Medical Sciences*, January 2011. http://www.acmedsci.ac.uk/index.php?pid=47&prid=88

'These restrictions mean that once the charity Cancer Research UK has funded a trial, it takes an average of 621 days before the first patient is treated': CRUK submission to AMS inquiry, June 2010, p.13. http://info.cancerresearchuk.org/prod_consump /groups/cr_common/@nre/ @pol/documents /generalcontent/cr_053410.pdf

'Another European initiative, the Physical Agents Directive, would have banned much medical use of magnetic resonance imaging scans, had not scientists eventually lobbied the European Commission to see sense': for more details, see Sense About Science's report at http://www.senseaboutscience.org/pages/physical-agents-directive.html

'"It's a bit like the decline and fall of the Roman Empire," says Sir Mark Walport': interview with author, 21/3/2011.

'The 2011 Bateson report into primate research noted': review of Research Using Non-Human Primates, Sir Patrick Bateson et al., July 2011. Report available at: http://blogs.nature.com/news/Bateson_Report.pdf See also Hannah Devlin and Mark Henderson, *The Times*, 'Monkey welfare standards "hamper medical research"': 28/7/2011. http://www.thetimes.co.uk/tto/science/article3107630.ece

'Professor Colin Blakemore, an Oxford University neuroscientist who was a vocal defender of such work, was attacked and letter bombs were addressed to his children': I am grateful to Colin Blakemore for his observations on animal extremism and political response. Interview, 26/7/2011.

'In 2000, the governing Labour Party caved in to activist pressure and cravenly sold its pension fund's shares in Huntingdon Life Sciences': 'Huntingdon Life Sciences stage rebound', *London Evening Standard*, 9/2/2000. http://www.thisismoney.co.uk/money/news/article-1573640/Huntingdon-Life-Sciences-stage-rebound.html

'In 2002, Laura Cowell, a sixteen-year-old girl with cystic fibrosis, agreed to front a counter-campaign extolling the medical benefits of animal experimentation': 'I would be dead a long time ago', Jon Boone, *Financial Times*, 29/7/2004. Laura Cowell died of her disease in 2011, while I was writing this book. I remain in awe of her bravery.

'Another teenager, Laurie Pycroft, formed a group called Pro-Test to support the Oxford lab, which staged a march of more than 1,000 people through the city in 2006. "I'm sick of seeing only the anti-vivs' argument being represented," he told the rally. "It's time to speak out in support of scientific research" ': 'Teenager leads fightback against anti-vivisectionists', David Harrison and Nina Goswami, *Daily Telegraph*, 26/2/2006. http://www.telegraph.co.uk/news/uknews/1511509/Teenager-leads-fight-back-against-anti-vivisectionists.html

'As more and more scientists took time to explain the case for animal experimentation, public support increased: the proportion not concerned about vivisection, or who are prepared to accept it when there are no alternatives, almost doubled from 32 per cent in 1999 to 60 per cent in 2009': IPSOS-MORI poll, 'Views on Animal Experimentation', March 2011. http://www.ipsos-mori.com/Assets/Docs/SRI_HEALTH_BIS_NC3Rs_combined_animal_experimentation_2011_FINAL_report_PUBLIC_110411.pdf

'Nicola Blackwood, has made assiduous efforts to court science and scientists since her narrow victory': for example by participating in the Royal Society pairing scheme; see *Times Eureka*, 7/7/2011, 'When Science and Politics Collide'. http://www.thetimes.co.uk/tto/science/eureka/article3085661.ece
    I don't want to suggest here, by the way, that heading off a geek campaign against her is the only reason she has taken an interest in science.

'Just before the 2010 general election, James O'Malley and Craig Lucas set up a website http://skeptical-voter.org

'During the 2010 election campaign, Martin Robbins, who writes the *Lay Scientist* blog for the *Guardian*, sent detailed questions to each of the main parties': 'The Litmus Test: Science policy at the election', Robbins, *Guardian*, 5/5/2010. http://www.guardian.co.uk/science/2010/may/05/science-policy-labour

'The Campaign for Science and Engineering also solicited and published letters from the party leaders about their plans for science': the letters are available on the CaSE website: http://sciencecampaign.org.uk/?page_id=659

'Daniel Sarewitz, Professor of Science and Society at Arizona State University, who writes a political column for the journal *Nature*, has argued that excess politicisation of this sort has been a significant factor in the damaging breakdown in relations between science and the Republican party in the US': '*Lab Politics*, "Most scientists in this country are Democrats. That's a problem" ': Daniel Sarewitz, 8/12/2010. http://www.slate.com/articles/health_and_ science/science/2010/12/lab_politics.html

'So when Michael Brooks decided to stand against him, he knew he wasn't about to win': details from interview with author, 9/2/2011.

' "There's a cultural problem here, in that scientists often see getting involved in politics, getting involved in government in general in fact, as pretty dirty," says Bill Foster': interview with author, 18/2/2011.

'John Denham, one of the few recent Cabinet ministers with a science degree, agrees. "It's always a good idea to have people with a relevant background," he says': interview with author, 7/3/2011.

' "If you could get maybe 25 per cent with a background in science and technology, their expertise would diffuse into the whole culture of Congress," Foster says': interview with author, 18/2/2011.

' "I went straight from science to Congress, but that's atypical," says Foster. "You want people with a record, but few scientists have even stood for their school board" ': interview with author, 18/2/2011.

'The constituency party meetings that choose their parliamentary candidates are often attended by a hundred or so members and are settled by majorities of little more than a dozen': see for example 'Fairer for Voter Numbers. Worse for the Voters', Daniel Finkelstein, *The Times*, 14/9/2011. http://www.thetimes.co.uk/tto/opinion /columnists /danielfinkelstein/article3163405.ece

' "There's really no alternative to engaging in electoral politics," says Denham': interview with author, 7/3/2011.

'In 2011, he joined another physicist who made it to Congress, the Republican Vern Ehlers, to launch Ben Franklin's List': Foster, interview with author, 18/2/2011. See also 'Groups Call for Scientists to Engage the Body Politic', Cornelia Dean, *New York Times*, 8/8/2011. http://www.nytimes.com/2011/08/ 09/science/09emily.html?_r=2 I'm also grateful to Eli Kintisch, of *Science*, for introducing me to Congressman Foster and alerting me to his initiative through his writing.

' "An Emily's List for science could have a very negative effect on the likes of Phil Willis, non-scientists who are good on science," says James Wilsdon, director of policy at the Royal Society': interview with author, 21/1/11.

' "They always know that it guarantees freedom of religion, speech and assembly," [Sean Tipton] says': interview with author, 21/2/11.

'Jon Spiers, head of public affairs at Cancer Research UK, says his team regularly meets with bemusement when it consults researchers funded by the charity about their policy concerns': interview with author, 2/3/2011.

'Their response was not just to complain, though they did use the media rather effectively to highlight both their concerns and those of the patients their work was ultimately intended to help': The campaign is recounted in '*Hype, hope* and *hybrids*. Science, policy and media perspectives of the. Human Fertilisation and Embryology Bill', edited by Dr Geoff Watts. http://www.acmedsci.ac.uk/download.php?file= /images/publication/Embryosr.pdf

'Gordon Brown, the Prime Minister, wrote a supportive opinion piece for the *Observer*', 18/5/2008. Why I believe stem cell researchers deserve our backing http://www.guardian.co.uk/commentisfree/2008/may/18/stemcells.medical research

'Caroline Flint, the Health Minister in charge, had yet to commit to a ban, making it easier for her to change her mind': 'Keep politics out of the lab', Mark Henderson, *The Times*, 4/6/2007; I made a point of praising her for her sensible U-turn. http://www.timesonline.co.uk/tol/comment/columnists/guest_contributors/article1878654.ece

'Nicola Blackwood, the new Oxford West and Abingdon MP, says that an overflowing inbox has a particular impact when it concerns a matter with which she's not especially familiar': interview with author, 11/2/2011.

' "You need to be the campaign group that gives ministers the biggest headache," said Tracey Brown, of Sense About Science. "That way, they can't ignore you" ': interview with author, 13/1/2011.

'In the winter of 2010–11, she was among sixteen MPs and twelve civil servants who took part in the pairing scheme organized by the Royal Society': the full list of participants is available at http://royalsociety.org/training/pairing-scheme/previous-participants

' "Most scientists feel sort of helpless before the political process, like leaves thrown in the wind," Wark says': interview with author, 11/2/2011.

' "My MP is Justine Greening, who's a Treasury minister, and you know I don't think I've ever actually met her," admits Mark Walport': interview with author, 21/3/2011.

'Mary Woolley, President of Research! America, a US advocacy group, describes the challenge as "the Starbucks test" ': 'Winning over the Republicans', Peter Aldhous, *New Scientist*, 4/1/2011. http://www.newscientist.com/article/dn19913-winning-over-the-republicans.html

3 Policy-based Evidence

' "What he said was that closing down special needs schools and putting needy kids into mainstream education is a lousy idea," said Hugh Abbott, the Secretary of State for Social Affairs and Citizenship': exchange from *The Thick of It*, series 2 episode 3.

'. . . the new Prime Minister decided that one way to set him apart from his long-term rival would be to reverse the government's recent decision to reclassify cannabis as a class C drug': see for example 'Brown announces review of cannabis classification', Will Woodward, *Guardian*, 19/7/2007. http://www.guardian.co.uk/politics/2007/jul/19/drugsandalcohol.immigrationpolicy

'His Home Secretary, Jacqui Smith, ordered the ACMD to reconsider': 'Gordon Brown planning clampdown on cannabis over health concerns', Francis Elliott and Richard Ford, *The Times*, 9/1/2008. http://www.timesonline.co.uk/tol/news/politics/article3156255.ece

'It found that the evidence, as it had done in 2004, supported the classification of cannabis in class C, and twenty of its twenty-three members voted against a change in the law': ACMD report, 'Cannabis: classification and public health', 2008. http://www.homeoffice.gov.uk/publications/alcohol-drugs/drugs/acmd1/acmd-cannabis-report-2008

'Sir Michael Rawlins, its chairman, said that reclassification "is neither warranted, nor will it achieve its desired effect": 'Scientists warn Smith over cannabis reclassification', Alan Travis, *Guardian*, 8/5/2008. http://www.guardian.co.uk/politics/2008/may/08/drugspolicy.drugsandalcohol

'They sought out other, more congenial experts whose research suggested a stronger link than had hitherto been appreciated between cannabis and psychosis': such research includes 'Cannabis use and risk of psychotic or affective mental health outcomes: a systematic review', Moore, Zammit, Lingford-Hughes, Barnes, Jones, Burke and Lewis, *Lancet*, Volume 370, Issue 9584, pp. 319–328, 28/7/2007. http://www.thelancet.com/journals/lancet/article/PIIS0140-6736% 2807%2961162-3/abstract. The merits of this work are beyond the scope of the discussion here. The point is that it was weighed by the ACMD, which found it insufficient to warrant reclassification as class B.

' "There is a compelling case to act now rather than risk the future health of young people," she said': 'Cannabis to be reclassified as a class B drug', Christopher Hope, *Daily Telegraph*, 7/4/2008. http://www.telegraph.co.uk/news/1934756/Cannabis-to-be-reclassified-as-a-class-B-drug.html

Brown and cannabis: http://www.timesonline.co.uk/tol/news/politics/article3156255.ece

'Britain's death rate from heart attacks, for example, was double that in France': 'Mortality: Ministerial briefing on NHS reforms', available online at http://prestonlibdems.org.uk/en/article/2011/462883/modernising-the-nhs-the-health-and-social-care-bill

'There was, he said, "a range of evidence" that GP commissioning was the best way to improve matters': BMA web interview, 7/3/2011. http:// www.bma.org.uk/ healthcare_policy/nhs_white_paper/nhsreformlansleylive.jsp

'John Appleby, the chief economist of the King's Fund, a health think-tank, began to look carefully at the mortality statistic Lansley was quoting': 'Does poor health justify NHS reform?', John Appleby, *British Medical Journal*, 27/1/2011. http://www.bmj.com/content/342/bmj.d566.full

'when Ben Goldacre, the *Guardian*'s "Bad Science" columnist, put it under his microscope': 'Andrew Lansley and his imaginary evidence', Ben Goldacre, *Guardian*, 5/2/2011. http://www.badscience.net/2011/02/andrew-lansley-and-his-imaginary-evidence/ Papers quoted by Goldacre: Kay: http://www.ncbi.nlm.nih.gov /pmc/articles/PMC1314221/ Greening and Mannion: http://www.bmj.com/ content /333/7579/1168.full Petchley: http://linkinghub.elsevier.com /retrieve/pii/ S0140673695918051 Coulter: http://eurpub.oxfordjournals.org/content/5/4/233 .abstract

'then added what David Halpern, a senior adviser to both Tony Blair and David Cameron, has called "spray-on evidence" ': interview with Tim Harford, *More or Less*, BBC Radio 4, 4/6/2010. http://bbc.in/foZ2q0

'They are practised exponents of what Richard Feynman famously christened "cargo-cult science" ': Feynman, 'Cargo-Cult Science', commencement address to CalTech, 1974; reproduced in Feynman, *Surely You're Joking, Mr. Feynman!* (W. W. Norton, 1985).

'Imaginary evidence': Ben Goldacre used this phrase first to describe Andrew Lansley's approach; see reference above.

'When one of the council's more eminent members, the future Nobel laureate Elizabeth Blackburn, began to dissent vocally from its opinions, she was sacked':

'Bush Ejects Two from Bioethics Council', Rick Weiss, *Washington Post*, 28/2/2004. http://www.washingtonpost.com/ac2/wp-dyn?pagename=article&contentId= A13606-2004Feb27 For a good account of the Elizabeth Blackburn affair, see also Chris Mooney, *The Republican War on Science*, pp. 198–204.

'Clairvoyant evidence': thanks to Mark Stevenson for the phrase.

'In May 2006, Patricia Hewitt, a Labour predecessor of Lansley's as Health Secretary, briefed the *Independent on Sunday*': 'Childbirth revolution: Mummy State', Marie Woolf and Sophie Goodchild, *Independent on Sunday*, 14/5/2006. http://www. independent. co.uk/life-style/health-and-families/health-news/childbirth-revolution-mummy-state-478174.html

'the following year, the Royal College of Obstetricians and Gynaecologists issued evidence-based guidelines that endorsed home birth': for example, RCOG / Royal College of Midwives guidance, 2007. http://www.rcog.org.uk/womenshealth/clinical-guidance/home-births

'"at best a peripheral policy concern, and at worst a political bargaining chip"': 'Putting Science and Engineering at the Heart of Government Policy'. The Science and Technology Committee was then called the Innovation, Universities, Science and Skills Committee. http://www.parliament.the-stationery-office.co.uk/pa/cm200809 /cmselect/cmdius/168/168i.pdf

'Bovine tuberculosis is one of the most serious agricultural diseases in Britain: it led to the slaughter of 25,000 cattle in 2010, and cost the taxpayer £90 million': Defra statistics published at www.defra.gov.uk/animal-diseases/a-z/ bovine-tb

'evidence from a major, high-quality study commissioned by the government: the Randomized Badger Culling Trial': full report at http://collections.europarchive. org/tna/20081027092120/http:/defra.gov.uk/animalh/tb/culling/index.htm

For a good account of the problems with the badger cull, see 'How the badger got stuck in the middle', Samantha Weinberg, *The Times*, *Eureka*, 6/1/2011. http://www.thetimes.co.uk /tto/science/ eureka /article2862421.ece And: Letter from Lord Krebs to *The Times*, 12/7/2011. www.thetimes.co.uk/tto/opinion/letters/ article3090704.ece

' "Where is it?" asked Imran Khan': 'Goodbye to golden hellos for science teachers', Imran Khan, 1/2/2011. http://sciencecampaign.org.uk/?p=2792

'Tracey Brown, who campaigns for greater use of evidence in public policy as direc- tor of Sense About Science, is clear on its limitations': interview with author, 13/1/2011.

'Evan Harris, the former Liberal Democrat MP who has set up the Centre for Evidence-Based Policy, agrees': interview with author, 23/12/2010.

'In his book *The March of Unreason*': Dick Taverne, *The March of Unreason*, OUP 2005. http://www.amazon.co.uk/March-Unreason-Science-Democracy-Fundamentalism/ dp/0192804855

'As Tim Harford describes in his book *Adapt*': Tim Harford, *Adapt*, Little, Brown, 2011, chapter 2. http://www.amazon.co.uk/Adapt-Success-Always-Starts-Failure/dp/ 1408701529/ref=sr_1_1?s=books&ie=UTF8&qid=1318280158&sr=1-1

' "I don't think there are very many scientists who are naïve enough to think that science should always determine outcomes," said John Holdren, before becoming President Obama's chief scientific adviser. "But you shouldn't defend outcomes by distorting the science" ': quoted in Mooney, *The Republican War on Science*, p. 23.

'Iceland, the world's eighteenth largest island, sits in the North Atlantic at the junction between the Eurasian and North American plates': most of the details that follow are from the Commons Science and Technology Committee report, 'Scientific advice and evidence in emergencies', 2011. http://www.publications.parliament.uk/pa/cm201011 /cmselect/cmsctech/498/498.pdf

'Volcanologists and air-traffic controllers': Eric Campbell, ABC News, 25/5/2010. http://www.abc.net.au/foreign/content/2010/s2909102.htm

' "Half of all business ideas fail, and 10 per cent of companies fail every year," Tim Harford points out. "Why should we assume that politics does any better?" ': interview with author, 25/1/2011.

'Medical research is indexed by an online archive called PubMed, and the Cochrane and Campbell Collaborations': web addresses are http://www.ncbi.nlm.nih.gov/pubmed/ www.cochrane.org www.campbellcollaboration.org

'Web-based initiatives such as Straight Statistics, FullFact.org and Channel 4 News's FactCheck': web addresses are http://fullfact.org/http://blogs.channel4.com/ factcheck/www.straightstatistics.org

'Straight Statistics, for example, has demolished a string of misleading government announcements about crime rates': see for example 'A blunt way to cut knife crime', Nigel Hawkes, 22/07/2009. http://www.straightstatistics.org/article/blunt-way-cut-knife-crime

'Evidence abuse is a form of what the journalist Peter Oborne calls political lying, and politicians never look good when their lies are found out': Peter Oborne, *The Rise of Political Lying*, Free Press, 2005.

' "You should see your MP as somebody who gives you the ability to hold bigger and more powerful people than yourself to account," says Tracey Brown. "Bringing issues to their attention isn't necessarily lobbying. You can help them out, feed them information they can use": interview with author, 13/1/2011.

'When the Department of Health ran a consultation before drafting the Human Fertilisation and Embryology Bill, it received 535 responses': 'Review of the Human Fertilisation and Embryology Act: proposals for revised legislation (including establishment of the Regulatory Authority for Tissue and Embryos)', 14/12/2006. http://www.dh.gov.uk/en/Publicationsandstatistics/Publications/Publications PolicyAndGuidance/DH_073098

'Dave Wark, the physicist we met in the last chapter, started to contribute to select committees only after taking part in the Royal Society pairing scheme, testifying on physics funding and student visas': interview with author, 11/2/2011.

'CaSE has only 925 individual members': email from Imran Khan, director of CaSE.

'[Science is Vital] emailed its subscribers inviting them to share their concerns about the structure of science careers with David Willetts, the Science Minister': its report, 'Careering out of Control', was published on 6/10/2011. http://scienceisvital .org.uk/2011/10/06/careering-out-of-control-a-crisis-in-the-uk-science-profession

'Evan Harris hopes his Centre for Evidence-Based Policy': interview with author, 23/12/2010.

'[NESTA], was establishing a UK Alliance for Useful Evidence' :http://www.nesta.org.uk/ events/assets/events/building_the_uk_alliance_for_useful_evidence

' "We need to use the opportunities that events provide us with to make the use of evidence into a central theme of political debate," says Harris': Westminster Skeptics Geek Manifesto event, 7/2/2011.

'The Campaign for Science and Engineering has produced a "scorecard" that marks the influence of chief scientists according to six factors': unpublished at time of going to press. Web link will be published on the book's website.

'What Bob Watson, chief scientist at Defra, values most is his seat on the departmen-tal board': interview with author, 15/2/2011.

' "Chief scientific advisers need to have policy sign-off powers," ' says Chris Tyler, director of the Centre for Science and Policy at the University of Cambridge': inter-view with author, 24/1/2011.

' "The geek movement is often more comfortable outside the tent pissing in," says Tyler': interview with author, 24/1/2011.

'Professor Ben Martin, of the Science and Technology Policy Research Unit at the University of Sussex, argues that there's a need for more movement between the academic, industrial and civil service sectors': Centre for Science and Policy lecture, reported 5/3/2010. http://www.csap.cam.ac.uk/news/article-around-the-science-policy-world-in-50-mi

'The Centre for Science and Policy at Cambridge has begun a "Policy Fellowship" scheme' http://www.csap.cam.ac.uk/programmes/policy-fellowships

' "About 60 per cent of PhD students don't go on to do postdocs," says Tyler': interview with author, 24/1/2011.

'When George Osborne became Chancellor in 2010, one of his first decisions was to establish an Office for Budget Responsibility (OBR)': details available on the Treasury website. http://www.hm-treasury.gov.uk/d/press_01_10.pdf

'A system rather like this was in fact proposed by the Commons Science Committee, in its 2009 report, *Putting Science and Engineering at the Heart of Government Policy*': 'Putting Science and Engineering at the Heart of Government Policy'. http://www.parliament. the-stationery-office.co.uk /pa/cm200809 /cmselect/cmdius/168/168i.pdf. See also: 'Ministers who ignore science facts should be exposed, say MPs', Mark Henderson, *The Times*, 23/7/2009. http://www.thetimes.co.uk/tto/science /policy/article5470.ece

'Both Sir John Beddington and Bob Watson consider that having to conduct such a naming and shaming exercise would risk undermining their relationships with ministers and senior civil servants, diluting their day-to-day influence accordingly': interviews with author: Beddington, 1/2/2011; Watson, 15/2/2011.

'We can use carrots, offering our expertise freely and openly to those ministers, MPs and civil servants who are willing to accept it. But we also need sticks, effective ways of convincing politicians that abusing evidence carries a cost.' The Campaign for Science and Engineering and Sense About Science offers some practical advice on lobbying and engagement with policy and politics on its website. http://science campaign.org.uk/?page_id=6872

4 Unfair and Unbalanced

'Natalie Morton, an autopsy showed, was killed by a large chest tumour that had not been detected': 'Cervical cancer jab girl Natalie Morton died from large chest tumour', David Rose, *The Times*, 2/20/2009. http://www.timesonline.co.uk /tol/news/ uk/health/article6856774.ece

'"The sudden death of Coventry schoolgirl Natalie Morton after a jab against cervical cancer highlights the reality that vaccination programmes are not without their risks," wrote Richard Halvorsen, a GP critical of many vaccines, in the *Daily Mail*': 'I'm not opposed to jabs but there are serious worries', Dr Richard Halvorsen, *Daily Mail*, 1/10/2010. http://www.dailymail.co.uk/news/article-1217057/Dr-Richard-Halvorsen-Im-opposed-jabs-worries.html

'an inquest was hearing from Caron Grainger, director of public health for NHS Coventry': David Rose, cited above. http://www.timesonline.co.uk/tol/news/uk /health/article6856774.ece

'Doctors, the coroner and Natalie's parents all accepted that there was no link': 'Cancer girl "died of tumour"', BBC News Online. 1/10/2009. http://news. bbc.co.uk/1/hi/8284517.stm

'the *Today* programme on BBC Radio 4 held a discussion between Adam Finn, Professor of Paediatrics at the University of Bristol, and Halvorsen': 2/10/2009. http://news.bbc.co.uk/today/hi/today/newsid_8286000/8286445.stm

'a commitment to "due impartiality" is enshrined in the BBC Charter and its editorial guidelines': BBC Charter. http://www.bbc.co.uk/guidelines/editorialguidelines/ page/guidelines-impartiality-introduction

'As the science writer Chris Mooney put it': 'Blinded By Science: How "Balanced" Coverage Lets the Scientific Fringe Hijack Reality', Chris Mooney, *Columbia Journalism Review*, November 2004. http://comminfo.rutgers.edu/jri/hsj/Misguided Balance.pdf

'Steve Jones, the geneticist, reached a similar conclusion in a review of the BBC's science coverage he led in 2011': BBC Trust review of impartiality and accuracy of the BBC's coverage of science, July 2011, p. 58. http://www.bbc.co.uk/ bbctrust/assets/files/pdf/our_work /science_impartiality/science_impartiality.pdf

'The link was not even proposed in his infamous case series report in the *Lancet*; Wakefield raised it in a press conference, where his comments were unconstrained by the demands of peer-review': retracted paper, *Lancet*, Volume 351, Issue 9103, pp. 63 –41, 28/2/1998. RETRACTED: Ileal-lymphoid-nodular hyperplasia, non-specific colitis, and pervasive developmental disorder in children, AJ Wakefield et al. http://www.thelancet.com/journals/lancet/article/PIIS0140-6736%2897% 2911096-0/abstract

'As Tammy Boyce, of Cardiff University, describes in her book on the saga, *Health, Risk and News: the MMR Vaccine and the Media*, the impression given was of genuine dissension between experts': Tammy Boyce, *Health, Risk and News: the MMR Vaccine and the Media*, Peter Lang, 2007. http://books.google.com /books/about/Health_risk_and_news.html?id=2whS9tXKLn4C

Chapter 11 of Ben Goldacre's *Bad Science* also gives a good account of the media's response to what he calls the 'MMR hoax'.

'While some specialist health and science journalists, such as Nigel Hawkes in *The Times* and Sarah Boseley in the *Guardian*, robustly criticized Wakefield's claims': see for example: 'A lone doctor's fear set parents against experts – The MMR controversy', Nigel Hawkes, *The Times*, 7/2/2002; and 'Jab warning "wrong" ', Sarah Boseley, *Guardian*, 12/3/1998.

'one of Wakefield's most loyal newspaper cheerleaders, Lorraine Fraser of the *Sunday Telegraph*, was named Health Writer of the Year by her industry peers': an example of Fraser's work in the year she won the award was: 'Shame on officials who say MMR is safe', *Sunday Telegraph*, 21/1/2001. http://www.telegraph.co.uk/news/uknews/ 1318767/Shame-on-officials-who-say-MMR-is-safe.html. Fraser's award is recorded in the British Press Awards roll of honour http://www.pressgazette.co.uk/ hybrid.asp?typeCode=99

'In 1998–9, when the vaccine scare first surfaced, 88 per cent of two-year-olds in England had been immunized against measles, mumps and rubella; by 2003–4, coverage had fallen to 80 per cent, and as low as 61 per cent in some parts of London': statistics from the Health Protection Agency and Department of Health websites. Measles cases: http://www.hpa.org.uk/web/HPAweb&HPAwebStandard /HPAweb_C /1195733778332; Measles deaths: http://www.hpa.org.uk/web/ HPAweb&HPAwebStandard/HPAweb_C/1195733835814 1998–9 MMR coverage: http://www.dh.gov.uk/en/Publicationsandstatistics/Statistics/StatisticalWorkAreas /Statisticalhealthcare/DH_4080886 2003–4; MMR coverage: http://www.ic.nhs.uk /news-and-events/news/increase-in-mmr-vaccination-coverage-in-england- report-shows-but-child-immunisation-levels-are-still-lower-than-the-rest-of-the-uk 61 per cent in London: 'Measles-associated encephalitis in children with renal transplants: a predictable effect of waning herd immunity', Kidd et al., *Lancet* 2003; 362: 832. http://download.thelancet.com/pdfs/journals/lancet/PIIS0140 673603142560.pdf

'the *Daily Telegraph* ran a front-page splash under the headline: "New heart attack jab even more effective than statins"': Richard Alleyne, *Daily Telegraph*, 18/4/2011. http://www.telegraph.co.uk/health/healthnews/8459488/New-heart-attack-jab-even- more-effective-than-statins.html

'In his book *Bad Science*, Goldacre presents a catalogue of further ways in which the media misleads': Goldacre, *Bad Science*, Fourth Estate, 2008.

'In January 2011, Channel 4 News ran a scoop on pregnancies among women using a contraceptive implant called Implanon': http://www.channel4.com/news/implanon-con- traception-failures-cost-nhs-200-000. For criticism of the story, see Petra Boynton, 'Contraceptive Implants and Media Panics – what you need to know', 6/1/2011.http://www.drpetra.co.uk/blog/implanon-and-media-panics-%E2%80%93- what-you-need-to-know 'the media scare led thousands of women to abandon their contraception, and to 10,000 extra abortions': 'Abortions leapt after Pill scare', Jeremy Laurence, *Independent*, 1/10/1997. http://www.independent.co.uk/news/health-abortions-leapt-after-pill-scare-1233286.html

' "In their choice of stories, and the way they cover them, the media create a parody of science," Goldacre argues': Goldacre, *Bad Science*, p. 208.

'a 2009 study led by Andy Williams, of Cardiff University': 'Mapping the Field: Specialist science news journalism in the UK national media', Andy Williams, 2009. http://www.cardiff.ac.uk/jomec/research/researchgroups/riskscienceandhealth/funded projects/mappingscience.html

'*The Times*, as we have seen, launched *Eureka*, a glossy monthly science magazine, in 2009': published on first Thursday of month; available online at www.thetimes.co.uk/eureka

'members of the BBC's award-winning Natural History Unit complained to Steve Jones that they were warned to avoid too much about ecology or evolution, so they made "children's programmes, albeit superbly good children's programmes"': BBC Trust review of impartiality and accuracy of the BBC's coverage of science, July 2011, p. 38. http://www.bbc.co.uk/bbctrust/assets/files/pdf/our_work/science_impartiality/science_impartiality.pdf

'Many a health scare has begun when a tabloid executive fell sick and generalized from his own perceived experience, or concluded that there must be an epidemic': I have been told that a newspaper's breathless coverage of a winter vomiting outbreak began when a senior executive fell sick with it over Christmas and concluded there must be a major epidemic. I have pledged to keep this source anonymous.

'When Ian Plimer, the author of a book criticizing the science of climate change, appeared on *Today* in 2009, he made a string of extraordinary claims': the interview, on 12/11/2009, can be heard at http://news.bbc.co.uk/today/hi/today /newsid_8356000/8356369.stm. For criticism, see George Monbiot, 'Should climate deniers be allowed to speak on the *Today* programme?', *Guardian* blog, 13/11/2009. http://www.guardian.co.uk/environment/blog/2009/nov/13/climate-deniers-today-programme

'In his review of BBC science coverage, Steve Jones singled out another interview by Webb, with a campaigner against GM crops called Kirtana Chandrasekaran, for particular criticism': the interview, conducted on 8 June 2010, can be heard at http://news.bbc.co.uk/today/hi/today/newsid_8727000/8727561.stm The criticism comes from the Jones report for BBC Trust, cited above, p. 62.

'Even Evan Davis, the former BBC economics correspondent who is the *Today* presenter most comfortable with science, is not immune to the odd howler': *Today* programme, 12/3/2011.

'Simon Jenkins in the *Guardian* likes to repeat the charge that science is too big for its boots: see for example 'Martin Rees makes a religion out of science so his bishops can gather their tithe', *Guardian*, 24/6/2010. http://www.guardian.co.uk/commentisfree/2010/jun/24/rees-makes-religion-out-of-science

'Just one scientist, Brian Cox, was invited to appear on *Question Time* throughout 2010 – the BBC's "Year of Science"': the BBC does not publish a full list of guests, but several online sources have kept a record. http://en.wikipedia.org/wiki/List_of_Question_Time_episodes http://www.medialens.org /forum/viewtopic.php?t=2722

'This oversight speaks volumes about media attitudes. It is not as if there is a shortage of scientists with the intellectual breadth to contribute to political debate: Sir Paul Nurse, Lord Rees of Ludlow, Dame Nancy Rothwell and Professor Colin Blakemore are just a few big names with interesting perspectives': just before this book went to press, on 10 November 2011, Colin Blakemore finally appeared on the programme.

'Trust in the media is on the wane': see for example 'Number cruncher: a matter of trust', Peter Kellner, *Prospect*, 22/9/2010. http://www.prospectmagazine.co.uk/2010/09/peter-kellner-yougov-trust-journalists

'The 2011 Public Attitudes to Science report, commissioned from Ipsos-MORI by the UK Government Office for Science': 'Public Attitudes to Science', Ipsos-MORI, May 2011. http://www.ipsos-mori.com/Assets/Docs/Polls/sri-pas-2011-main-report.pdf p3

'[Harding] often notes that in 2008, there were three stories that reliably increased circulation when they appeared on the front page. One was Barack Obama. The second was the global financial crisis. The third was the Large Hadron Collider': Speech to World Congress of Science Journalists, 2/7/2009.

'In August 2011, the science section of [the *Guardian*'s] website attracted almost 4 million page impressions and 2.8 million unique users': email from Alok Jha, *Guardian* science correspondent, 23/9/2011.

'The *Sunday Times* followed up with a scoop of its own': 'UN climate panel shamed by bogus rainforest claim; An IPCC report warning of Amazon catastrophe was unfounded', Jonathan Leake, *Sunday Times*, 31/1/2010. Original article available online at http://www.realclimate.org/docs/Leake_and_North_original_S_Times_article_31_Jan_2010.pdf. Retracted from *Sunday Times* website.

'Lewis, however, had told the paper that the IPCC's statement was "poorly written and bizarrely referenced, but basically correct"': quoted in 'Forests expert officially complains about "distorted" *Sunday Times* article', David Adam, *Guardian*, 24/3/2010. http://www.guardian.co.uk/environment/2010/mar/24/sunday-times-ipcc-amazon-rainforest

'The watchdog sided with the scientist, forcing the *Sunday Times* into an apology': http://www.thesundaytimes.co.uk/sto/news/article196428.ece. See also University of Leeds press release: http://www.leeds.ac.uk/news/article/839/leading_scientist_forces_climate_article_apology_ and_retraction

'"In China, a young woman is having open heart surgery," said the presenter': *Alternative Medicine: The Evidence*, broadcast 24/1/2006; quoted in 'A groundbreaking experiment … or a sensationalised TV stunt?', Simon Singh, *Guardian*, 25/3/2006. http://www.guardian.co.uk/media/2006/mar/25/science.broadcasting

'Simon Singh complained to the BBC Trust, which upheld his criticism': BBC Trust: Editorial Complaints, April 2007 (issued May 2007), p. 5. http://www.bbc.co.uk/bbctrust/assets/files/pdf/appeals/esc_bulletins/apr2007.pdf

'"I found it quite shocking that the BBC did that and two million people were misled," says Singh': interview with author, 29/12/2010.

'the *Guardian* blogger Martin Robbins wrote an inspired spoof under the headline "This is a news website article about a scientific paper"': *Guardian, Lay Science* blog, 24/9/2010. http://www.guardian.co.uk/science/the-lay-scientist/2010/sep/24/1

'It was recommended 38,000 times on Facebook, it was retweeted 4,800 times on Twitter, and it received 650,000 page views in just a few days. On the day it was published, it accounted for about 15 per cent of the *Guardian* website's entire traffic': email from Martin Robbins, 16/9/11.

'In May 2010, the journal *Nature Neuroscience* published a paper about acupuncture from a team at the University of Rochester, which purported to demonstrate a painkilling effect on mice': 'Adenosine A1 receptors mediate local anti-nociceptive effects of acupuncture', Goldman, Nedergaard et al., *Nature Neuroscience* 13, 883–88 (2010). http://www.nature.com/neuro/journal/v13/n7/full/nn.2562.html

' "Let's get straight to the point: acupuncture DOES ease pain," said the *Daily Mail*'s headline':, Fiona Macrae, *Daily Mail*, 31/5/2010. http://www.dailymail. co.uk/health/article-1282678/Acupuncture-DOES-ease-pain.html. See also the press release, which shows how journalists aren't always responsible for the misinterpretations – even if they often lap them up. http://www.eurekalert.org/pub_releases/ 2010-05/uorm-ame052710.php

'Where the mainstream media failed, bloggers such as Ed Yong and David Gorski stepped in': Ed Yong, 'A biological basis for acupuncture, or more evidence for a placebo effect?', *Not Exactly Rocket Science* blog, 30/5/2010. http://blogs. discovermagazine.com/notrocketscience/2010/05/30/a-biological-basis-for-acupuncture-or-more-evidence-for-a-placebo-effect; David Gorski, 'Another overhyped acupuncture study misinterpreted', *Science Based Medicine* blog, 2/6/2010. http:// www.sciencebasedmedicine.org/index.php/another-overhyped-acupuncture-study-misinterpreted

'The charity Cancer Research UK, for example, runs an outstanding daily blog that picks apart the science behind news stories about cancer, and explains major advances in the field': see for example http://scienceblog.cancerresearchuk.org /2011/09/12/ crocus-smart-bomb-cancer-cure-its-a-bit- more-complicated-than-that

'*Behind the Headlines*, a website run by NHS Choices, does a similar job for a still wider range of health stories' http://www.nhs.uk/News/Pages/NewsIndex.aspx

'As Abraham Lincoln said, "If you would win a man to your cause, first convince him that you are his sincere friend" ': Temperance Address, Springfield, Illinois, 22/2/1842. http://showcase.netins.net/web/creative/lincoln/speeches/temperance.htm I'm glad to have found this reference, because of another great quote attributed to Lincoln: 'The problem with quotations on the internet is you can never be sure they're authentic.'

'I picked up on a story in *New Scientist* about the forthcoming shut-down of the Large Electron-Positron Collider at the CERN particle physics lab near Geneva': 'No sign of the Higgs boson', Eugenie Samuel, *New Scientist*, 5/12/2001. http://www.newscientist. com/article/dn1649-no-sign-of-the-higgs-boson.html

'My story strongly implied that the search for it had been a waste of money': 'God particle disappears down £6bn drain', Mark Henderson, *The Times*, 6/12/2001.

'At the height of the false controversy over MMR in 2003, Channel Five broadcast a hagiographic dramatisation of Andrew Wakefield's campaign against the vaccine': *MMR: Hear the Silence*, broadcast 15/12/2003. For criticism of the debate, see Michael Fitzpatrick, 'MMR: fact and fiction', Spiked Online, 11/12/2003. http://www.spiked-online.com/articles/00000006E00D.htm

' "Scientists should be good at doing science, just as my bank manager should be good at managing a bank," says Simon Singh': interview with author, 29/12/2010.

' "You can't move in some physics departments without people slagging off Brian Cox," ' says Mark Stevenson, a comedian and science writer': interview with author, 2/2/2011.

' "For scientists who would be agents of change, communication is not an add-on," says Nancy Baron, director of science outreach at the marine biology group Compass. "It is central to their enterprise" ': 'Stand up for science', Nancy Baron, *Nature*, 468, 19/12/2010, 1032–33. http://www.nature.com/nature/journal/v468/n7327 /full/4681032a.html

' "Between two early press briefings and the final vote in Parliament, the scientists involved had talked to journalists hundreds of times and done scores of media interviews," recalled Fiona Fox, director of the Science Media Centre, in an Academy of Medical Sciences pamphlet published when the Bill became law': '*Hype, hope and hybrids*. Science, policy and media perspectives of the Human Fertilisation and Embryology Bill', edited by Dr Geoff Watts, p. 11. http://www.acmedsci.ac.uk /index.php?pid=101&puid=151

5 Geekonomics

'When the Conservative minister William Waldegrave cleared his desk as Chief Secretary to the Treasury following Labour's landslide election victory of 1997, he left a short note for his successor, Alistair Darling': press conference for launch of Royal Society Scientific Century report, 8/3/2010; reported in 'Royal Society says research funding cuts will lead to economic decline', Mark Henderson, *The Times*, 9/3/2010. www.timesonline.co.uk/tol/news/science/article7054472.ece

'Labour's outgoing Chief Secretary, Liam Byrne, left a less helpful note for his successor, David Laws': 'Labour minister Liam Byrne left note on desk: "There's no money left"', Suzy Jagger, *The Times*, 17/5/2010. http://business.timesonline.co.uk /tol/business/economics/article7128665.ece

'Adam Afriyie, the Tories' science spokesman in opposition, had made it plain that "major science budget cuts are inevitable" ': 'Adam Afriyie: From Peckham council house to shadow minister', David Cohen, *London Evening Standard*, 8/2/2010. http://www.thisislondon.co.uk/standard/article-23803233-adam-afriyie-from-peckham-council-house-to-shadow-minister.do

'The gravity of the situation was underlined on 8 September, when the Cabinet minister responsible for science – Vince Cable, the Business Secretary – delivered his first speech on the subject': Vince Cable, speech at Queen Mary's, University of London, 8/9/2010. http://www.bis.gov.uk/news/speeches/vince-cable-science-research-and-innovation-speech

'He added, erroneously, that 45 per cent of public funding rewards research that is "not of an excellent standard" ': BBC Radio 4 *Today* programme, 8/9/2010. http: //news.bbc.co.uk/today/hi/today/newsid_8978000/8978968.stm For more detailed criticisms of Cable's claim, see 'Underrating our scientists isn't clever, Dr Cable', Mark Henderson, *The Times*, 9/9/2010; and 'How to read Vince Cable's speech on science at Queen Mary's', William Cullerne Bown, *Research Fortnight*, 8/9/2010. http://exquisitelife.researchresearch.com/exquisite_life/2010/09/how-to-read-vince-cables-speech-on-science-at-queen-marys.html

' "Sod it," she blogged. "Let's march on London!" ': 'In Which The Great Slumbering Scientific Beast Awakens', Jenny Rohn, *Mind the Gap* blog, 8/9/2010. http://blogs.nature. com/ue19877e8/2010/09/08/in-which-the-great-slumbering-scientific-beast-awakens

'The Science is Vital Campaign was born': for a good summary of the Science is Vital campaign, see Della Thomas's blogpost http://dellybeandiary.wordpress.com/ 2010/10/25/science-is-vital-the-journey

'Celebrities such as Brian Cox, Dara O'Brian and Patrick Moore offered enthusiastic backing, as did the journal *Nature* and scientific charities such as Cancer Research UK and the Wellcome Trust': notable signatories are listed at http://scienceisvital.org.uk/ 2010/09/30/signatures

'Huppert tabled an Early Day Motion endorsing Science is Vital that was signed by 141 MPs': EDM 767. www.parliament.uk/edm/2010-12/767

'Rohn was overwhelmed by the speed and scale of the reaction': interview with author, 25/1/2011.

'Instead of the 30 per cent reductions that had been pencilled in weeks earlier, the largest element of the science budget was to be frozen at £4.6 billion over the next four years': see 'Osborne gives science flat-cash reprieve for four years', Mark Henderson and Roland Watson, *The Times*, 19/10/2010. http://www.thetimes.co.uk/tto/science /article2773576.ece

' "Britain is a world leader in scientific research, and that is vital to our future economic success," he said': George Osborne, spending review statement, 20/10/2010. http://www.hm-treasury.gov.uk/spend_sr2010_speech.htm

'By adopting this message, long articulated by the Campaign for Science and Engineering, Science is Vital managed to speak in the mother tongue of the Treasury officials and Conservative politicians it needed to impress': Science is Vital, Key Messages. http://scienceisvital.org.uk/2010/09/28/key-messages

'In March 2010, the Royal Society had published *The Scientific Century*'. http://royalsociety.org/the-scientific-century

'On the same day, the engineering entrepreneur Sir James Dyson reached a similar conclusion in *Ingenious Britain*, a report commissioned by the Conservative Party': http://www.conservatives.com/News/News_stories/2010/03/Dyson_sets_out_plans_ to_boost_high_tech_industry.aspx

'The Prime Minister's Council on Science and Technology also endorsed the case for sustained investment, particularly given increased spending by rival nations such as the US, France, Germany and China': http://www.bis.gov.uk/assets/bispartners/ cst/docs/files/whats-new/10-584-vision-uk-research.pdf

'The Royal Society, often cautious in its discussions with ministers, explained in no uncertain terms that even a 10 per cent cut beyond inflation would start an unprecedented brain drain that would hit Britain's international competitiveness': Royal Society spending review submission, 12/7/2010. http://royalsociety.org/policy/ reports/spending-review-submission/

'A 2010 paper by Jonathan Haskel, Professor of Economics at Imperial College Business School, and Gavin Wallis, who is both a University College London academic and a Treasury official, was particularly influential': 'Public Support for Innovation, Intangible Investment and Productivity Growth in the UK Market Sector',

Haskel and Wallis, February 2010. http://spiral.imperial.ac.uk/bitstream/10044/1/5280/1/Haskel%202010-01.pdf

' "Current spend is around £3.5 billion," Haskel told the Commons Science and Technology Committee': http://www.parliament.the-stationery-office.co.uk/pa/cm200910/cmselect/cmsctech /335/335we66.htm. The size of the 'science budget' can be confusing. Haskel was quoting the science budget as calculated prior to the 2010 spending review. The spending review changed what it covered, to include research spending through universities and to remove capital spending. This explains the apparent discrepancy between Haskel's figure of £3.5 billion and Osborne's of £4.6 billion. For more information, see the Campaign for Science and Engineering's report 'Public funding of science and technology: putting government rhetoric to the test', 14/9/2011. http://www.sciencecampaign.org.uk/files/ScienceFundingSept2011.pdf

'The economic impact of one science sector was assessed in some detail in 2008 in *Medical Research: What's it Worth?*, a report commissioned by the Medical Research Council from the Health Economics Research Group, the Office of Health Economics, and RAND Europe': *Medical Research: What's it Worth?*, 2008. http://www.ohe.org/lib/liDownload/625/MRC%20Report%20Final.pdf

'Between 2003 and 2007, thirty-one university spin-out companies were floated on stock exchanges, with a combined value of £1.5 billion. In the same period, ten successful spin-out companies were acquired by larger businesses, for another £1.9 billion. Put together, they created some £3.4 billion for the UK economy – about as much as the research councils spend in a year': http://www.sciencecampaign.org.uk /documents/2010/CaSEjunebudgetbriefing2010.pdf

'A 2001 study by the Organisation for Economic Co-operation and Development found that increasing public spending on research and development raises productivity throughout the economy': http://ideas.repec.org/p/oec/stiaaa/2001-3-en.html http://www.oecd.org/dataoecd/26/32/1958639.pdf

'State- and charity-funded research also stimulates research in the private sector: in pharmaceuticals, an extra 1 per cent from taxpayers leads industry to spend an extra 1.7 per cent within eight years': 'Does public scientific research complement private investment in R&D in the pharmaceutical industry', Toole, *Journal of Law and Economics*, 2007, 50, 81–104.

'In June 2010, the Campaign for Science and Engineering persuaded senior executives from companies such as GlaxoSmithKline, Airbus and Syngenta to write to *The Times*': letter to *The Times*, 13/6/2010. http://www.timesonline.co.uk /tol/comment/letters/article7149421.ece

'The argument was: "Why is Rolls Royce here? Why is Glaxo here?," said one figure involved in the negotiations. "This was language the Treasury understood" ': quoted in Mark Henderson, *Eureka* daily post, 'According to George Osborne', Science is Vital, 20/10/2010. http://www.thetimes.co.uk/tto/science/eureka-daily/?blogId=Blog3dfc20db-8d88-49bd-9347-1957bc781c72Post34810042-f33b-4707-81ca-2bb4ce113d0b

'The House of Commons library estimates that total government spending on research in 2014-15 will be about 14 per cent lower in real terms than it was in 2010-11': Spending Review 2010; CaSE's Select Committee response, Nick Hall, 13/5/2011. http://sciencecampaign.org.uk/?p=5260 Bob Ward, *New Scientist*, 'Inflation erodes science budget', *The S Word* blog, 29/3/2011. http://www.newscientist.com/blogs/thesword/ 2011/03/inflation-erodes-uk-science-bu.html

'The Campaign for Science and Engineering has shown that when counted according to the previous Government's system, overall spending on science will fall from a planned £5.8 billion in 2010–11, to £5.4bn in 2014–15': 'Public funding of science and technology: putting government rhetoric to the test', Campaign for Science and Engineering, 14/9/2011. http://www.sciencecampaign.org.uk/files/ScienceFunding Sept2011.pdf

'Capital spending, on buildings, facilities and equipment, is being reduced significantly, by about 46 per cent over four years. The weak pound has put pressure on Britain's subscriptions to international science projects such as CERN, the European Space Agency and the European Southern Observatory, which are priced in stronger Swiss Francs or Euros': Spending Review 2010; CaSE's Select Committee response, Nick Hall, 13/5/2011. http://sciencecampaign.org.uk/?p=5260

'In 2009, a year before the cuts, a funding shortfall at the Science and Technology Facilities Council forced it to limit use of the ISIS neutron source, a key instrument for medical and environmental research, to just 120 days a year': 'Britain's budget woes continue', Geoff Brumfield, *Nature News* blog, 19/1/2011. http://blogs.nature. com/news/2011/01/britains_budget_woes_continue.html

'It also withdrew financial support from optical and infrared telescopes in the northern hemisphere, potentially denying British astronomers access': Commons Science and Technology Committee report, 'Astronomy and Particle Physios', 13/5/2011. http://www.publications.parliament.uk/pa /cm201012 /cmselect/cmsctech/806/806.pdf

'The UK, indeed, has spent a far smaller share of its national wealth on science than most of its leading economic rivals for two decades': figures from 'How is UK Science and Engineering Funded?', Campaign for Science and Engineering, 23/09/2010. http://www.sciencecampaign.org.uk/documents /2010/20100923CaSE briefingUKsciencefunding.pdf and: 'Securing our Economic Future with Science & Engineering', Campaign for Science and Engineering, June 2010. http://www.science campaign.org.uk/documents/2010/CaSEjunebudgetbriefing2010.pdf

'In 2010, France announced an extra €35 billion for research, while Germany pledged to increase spending on science and education by €12 billion by 2013': 'French research wins huge cash boost', Declan Butler, *Nature News*, 15/12/2009. http://www.nature.com/news/2009/091215/full/462838a.html

'The European Union's innovation "scoreboard", published in February 2011, showed the UK well behind its two biggest continental rivals, prompting Máire Geoghegan-Quinn, the European Commissioner for research, to warn Britain that it risks falling behind': Innovation scoreboard, 3//2/2011. http://www.eubusiness.com/topics /research /innovation-score.11

Máire Geoghegan-Quinn quoted in 'EU warning on support for research', Nikki Tait, *Financial Times*, 6/2/2011. http://www.ft.com/cms/s/0/4b9d6e20-323d-11e0-a820-00144feabdc0.html#axzz1LUefPZEh

'Australia increased spending on science by 25 per cent in 2009-10': 'Major increase in Australian science spending', Karen Harries-Rees, *Chemistry World*, 19/5/2009. http://www.rsc.org/chemistryworld/News/2009/May/19050901.asp

'Canada found an extra C$32 million for its three main research agencies in 2010–11': 'Science survives Canadian budget', Nicola Jones, *Nature News,* 9/3/09. http://www.nature.com/news/2010/100310/full/464153a.html

'In the US, President Obama's stimulus package included more than $100 billion for science, and his administration has been clear about its desire to enhance science funding on a more regular basis: the White House's 2011 Budget proposed a 5.9 per cent increase for research in an otherwise flat settlement': 'Obama goes "all in" for science', Peter Aldhous, *New Scientist*, 4/3/2009. http://www.newscientist.com/article /mg20126984. 000-obama-goes-all-in-for-science.html 'Obama budget backs basic science', Janet Fang et al., *Nature News*, 2/2/2011. http://www.nature.com/news /2010/100202 /full/ 463594a.html?s=news_rss

'That the Republican Congress forced Obama into an Osborne-like compromise, with funding for the National Institutes of Health and the National Science Foundation trimmed slightly, just goes to show that short-sighted politicians are not a species found uniquely in Britain's parliamentary habitat. The House of Representatives, indeed, had originally proposed much more swingeing cuts, including a $900 million raid on the Department of Energy's Office of Science': see 'Research Survives in 2011 Budget After Earlier Scare', Jeffrey Mervis, *Science Insider*, 12/4/2011. http://news.sciencemag.org/scienceinsider/2011/04/research-survives-in-2011-budget.html

At the time of writing, proposed cuts were 0.06 per cent for the NIH and 2.4 per cent for the NSF. See: '[Obama's] budget compromise agreed in summer 2011 seemed certain to cut the funds available for research': 'Senate Panel Trims NIH Budget By $190 Million', Jocelyn Kaiser, *Science Insider*, on 20/9/2011. http://news.sciencemag. org/scienceinsider/2011/09/senate-panel-trims-nih-budget-by.html 'Senate Panel Cuts NSF Budget by $162 Million', Jeffrey Mervis, *Science Insider*, 14/ 9/2011. http://news.sciencemag.org/scienceinsider/2011/09/senate-panel-cuts-nsf-budget-by.html

'Since 1999, China has increased its spending on science by almost 20 per cent every year, and now spends $100 billion a year on research: more than thirty times as much as the UK': 'China hints at science-funding reform', Hepeng Jia, *Nature News*, 21/3/2011. http://www.nature.com/news/2011/110321/full/ news.2011.173.html

'The UK used to have the same percentage target, until the present government cancelled it': Parliamentary answer by David Willetts, Science Minister, 27/4/2011. http://www.theyworkforyou.com/wrans/?id=2011-04-27b.52490.h

'A Royal Society report into international science, published in March 2011, noted that China is fast emerging as the next research superpower, which can soon be expected to match and even outstrip the achievements of traditional scientific nations': Royal Society report, 'Knowledge Networks and Nations'. http://royalsociety.org/policy/reports/knowledge-networks-nations

' "Quantity of input doesn't necessarily result in quality of output, but these investments are starting to yield results," explains James Wilsdon, the Royal Society's director of policy': Wilsdon, *Soapbox Science* blog, *Nature News*, 27/4/2011. http://blogs.nature.com /soapbox_science /2011/04/27/knowledge-networks-and-nations

'On current trends, China will begin to file more patent applications than the US, the current world leader, by 2015. "The impact of China's rise will depend largely on whether we are with them at the technology frontier or onlookers from the sidelines," the highly respected Institute for Fiscal Studies advised in September 2011. "We should choose the former" ': 'China is investing rapidly in skills and science: UK should do the same', Rachel Griffiths, Institute of Fiscal Studies, September 2011. http://www.ifs.org.uk/publications/5668

'While the company said the decision was no reflection on UK government policy, John Denham, Labour's Shadow Business Secretary at the time (and a chemistry graduate), has correctly identified that this is less reassuring that it seems': Campaign for Science and Engineering annual lecture, 9/3/2011. http://sciencecampaign. org.uk/?p=3918

'The Engineering and Physical Sciences Research Council (EPSRC), for instance, responded to the spending review by making plans to cut the number of PhD studentships it supports, potentially by more than a third, from 2,900 to 1,800': 'U.K. Slashes Science and Engineering Ph.D.s', Andy Extance, *Science Insider*, 15/8/2011. http://news.sciencemag.org/scienceinsider/2011/08/uk-slashes-science-and-engineeri.html

'"The internal logic in any bit of the system leads to cutting PhDs," says James Wilsdon. "But the aggregate of that would be a lost generation"': reported in Mark Henderson, 'Science and the spending review: why it matters and what lies ahead', *Eureka* daily blog, 18/10/2010. http://www.thetimes.co.uk/tto/science/eureka-daily/?blogId=Blog3dfc20db-8d88-49bd-9347-1957bc781c72Post56dca5fb-0102-43 5c-b11a-f04170de88ee

'When Sarah Palin, the darling of the Tea Party movement, ran for vice-president she highlighted fruit-fly research as just the sort of meaningless project that taxpayers shouldn't have to fund': Pittsburgh campaign speech, 24/10/2008.

'The specific project she criticized was spending just $211,000 on investigating a pest that threatens the $85 million California olive oil industry': Christopher Wanjek, *LiveScience*, 4/11/2008. http://www.livescience.com/5186-misdirected-criticism-palin-fruit-fly-remark.html

'John McCain, who chose Palin as his running mate, spent much of 2009 attacking science elements of the Obama stimulus package that he thought sounded silly, such as work on drug abuse and social behaviour': 'Stimulus Slammed: Republican Senators Release Report Alleging Waste', 3/8/2010. http://abcnews.go.com /GMA/ stimulus-slammed-republican-senators-release-report-alleging-waste/ story?id=11309090

'Darrell Issa, a Republican Congressman, took up the same baton in 2011 with a laundry list of research projects he deemed superfluous and wasteful': 'Issa Targets Silly-Sounding Research', *Wall Street Journal*; *Washington Wire* blog, 16/2/2011. http://blogs. wsj.com/washwire/2011/02/16/issa-targets-silly-sounding-research

'Mental illness costs the US at least $200 billion a year, and unplanned pregnancies another $5 billion': see for mental illness costs: http://www.time. com/time/health/article/0,8599,1738804,00.html and for unplanned pregnancy costs: http://www.thenationalcampaign.org/resources/pdf/FastFacts_DirectCosts_ UnplPreg.pdf

'A 2010 report by the Russell Group of leading UK universities, looking at the economic impact of 100 case studies of its institutions' research, found that it took an average of nine years for an initial discovery to yield a product licence, and another eight years after that for a spin-out company to turn a profit': 'The Impact of Research Conducted in Russell Group Universities', 2010. http://www. russellgroup.ac.uk/uploads/RG_ImpactOfResearch2.pdf

'Bill Foster, the physicist and former US congressman, points out that cutting science rarely has an immediate deleterious effect': interview with author, 18/2/2011.

'When Sir Tim Berners-Lee, a CERN computer scientist, developed the hypertext transfer protocol – the "http" of web addresses – he wasn't trying to invent a revolutionary form of mass communication that would transform countless businesses and enable the creation of entirely new ones': see 'CERN: Where the Web Was Born' http://public.web.cern.ch/public/en/about/web-en.html

'David Payne is not as well known as Berners-Lee, yet the Professor of Photonics at the University of Southampton can be just as fairly called a father of the modern internet': for a longer account of David Payne's work, see Mark Henderson, 'Science and the spending review: why it matters and what lies ahead', *Eureka* daily blog, 18/10/2010. http://www.thetimes.co.uk/tto/science/eureka-daily/?blogId= Blog3dfc20db-8d88-49bd-9347-1957bc781c72Post56dca5fb-0102-435c-b11 a-f04170de88ee

'Shankar Balasubramanian has a similar story to tell': see Kevin Davies, *The $1000 Genome*, pp. 86–91; and Royal Society of Chemistry, interview: 'The fundamentals of life', 17/4/2007. http://www.rsc.org/Publishing/Journals/cb/Volume/2007/5/Interview withShankarBalasubramanian.asp

'Andre Geim and Kostya Novoselov won the 2010 Nobel Prize for Physics for their discovery of grapheme': see Nobel prize press release. http://www.nobelprize. org/nobel_prizes/physics/laureates/2010/press.html

'Carl Sagan, the master of communicating by thought experiment, encapsulated the theme with his tale of the "Westminster Project" ': Sagan, *The Demon-Haunted World*, p. 384.

'As Simon Frantz, of the Nobel Foundation, puts it': Simon Frantz, email to author, 7/2/2011.

'In 2009, Lord Drayson, then the UK Science Minister, called for funding to be focused on areas of science in which Britain has a "strategic advantage"': 'Britain must specialize to make most of science expertise', Mark Henderson, *The Times*, 26/1/2009. http://www.thetimes.co.uk/tto/news/uk/article1967193. ece

'His government also introduced a new funding system for university research, which marks researchers according to the economic or social benefits of their work': 'Prove the benefits of research or lose funds, universities told', Greg Hurst and Mark Henderson, *The Times*, 12/11/2011. http://www.thetimes.co.uk/tto/education/article 2804906.ece

' "You don't know when you're starting where things are going to lead," [Langer] says': 'Robert Langer: UK science would be wrong to focus excessively on applied research', Mark Henderson, *Eureka* daily blog, *The Times*, 3/12/2009. http://www.thetimes.co.uk/tto/science/eureka-daily/?blogId=Blog3dfc20db-8d88- 49bd-9347-1957bc781c72Post9a048c08-8da2-46e2-bb6a-4abdc0b63018

'Simon Jenkins has argued that swine flu, BSE and many other health threats have been similarly exaggerated so scientists can feather their nests': 'Swine flu? A panic stoked in order to posture and spend', Simon Jenkins, *Guardian*, 29/4/2009. http://www.guardian.co.uk/commentisfree/2009/apr/29/swine-flu-mexico-uk-media1

'Rolf Heuer, the director general of CERN, betrayed this characteristic wariness in a 2011 article for *The Times* which otherwise made an impeccable case for the value of funding basic research': 'Fund the science that ends up in our pockets', Rolf Heuer, *The Times*, 2/2/2011. http://www.thetimes.co.uk/tto/opinion/columnists /article2896124.ece

'As the data presented by Science is Vital show (largely assembled by the Campaign for Science and Engineering)': http://scienceisvital.org.uk/2010/09/28/key-messages

'"I've made myself unpopular in the university sector for saying this, but there has to be an expectation that what it does will eventually turn into economic benefit, and that that should be demonstrated," Sir Mark Walport says. "We need to show why the government should be funding science, and how that funding delivers"': interview with author, 21/3/2011.

'Anna Grey, the research policy manager at the University of York, made a telling comment when the physics department took part in a pilot impact study': 'Turbulent ride for impact pilot', Anna Lewcock, *Chemistry World*, 28/6/2010. http://www.rsc.org/chemistryworld/News/2010/June/28061001.asp

'Pressed repeatedly by Senator John Pastore about its potential utility to national security, he replied': see Cornell University obituary, 2000. http://www. news.cornell. edu/releases/Jan00/RRWilson_obit.hrs.html

'"Science is going to have two battles in the years ahead," says John Denham': interview with author, 7/3/2011.

'From the right, it's been endorsed by George Will, the conservative *Washington Post* columnist who's drawn the ire of many scientists for his contrarian opinions about global warming': 'Needed: A science stimulus', George F. Will, *Washington Post*, 2/1/2011. http://www.washingtonpost.com/wp-dyn/content/article/2010/12/31/AR 2010123104129.html

'"Failing to invest sufficiently in science and skills can be short sighted," it said in a September 2011 report': 'China is investing rapidly in skills and science: UK should do the same', Rachel Griffiths, Institute of Fiscal Studies, September 2011. http://www.ifs.org.uk/publications/5668

'Science, says Denham, wants consistency and certainty in government policy. That's true for the wider economy as well': interview with author, 7/3/2011.

'The business leaders who wrote to *The Times* ahead of the spending review concurred': letter to *The Times*, 13/6/2010. http://www.timesonline.co.uk/tol/comment/letters/ article7149421.ece

6 Science Lessons

'The timetable at Monkseaton School on Tyneside sounds like a teenager's dream': see 'Roads to Enlightenment', Hannah Devlin and Mark Henderson, *The Times, Eureka*, 1/9/2011. http://www.thetimes.co.uk/tto/science/eureka/article3149602.ece

'Research led by Russell Foster, Professor of Circadian Neuroscience at the University of Oxford, has shown that their body clocks actually run several hours behind those of adults, because of differences in the timing hormone melatonin': 'The rhythm of rest and excess', Russell G. Foster and Katharina Wulff, *Nature Reviews Neuroscience*, Vol. 6, May 2005. http://www.nature.com/nrn/journal/v6/n5/abs/nrn

1670.html; 'The Young and Wise are Late to Rise', Russell G. Foster, *Times Higher Education Supplement*, 5/1/2007. http://www.timeshighereducation.co.uk/story.asp?storyCode=207366&sectioncode=26

'The National Association of Head Teachers described his idea as "totally inappropriate". John Dunford, the general secretary of the Association of School and College Leaders, said he knew from experience that the opposite was true, and that children work best first thing in the morning': 'Teachers stand up against "lie-in" plan for teenagers', *Guardian*, 9/3/09. http://www.guardian.co.uk/education/2009/mar/09/teenage-pupils-lie-in-plan

'In August 2011, after the first full school year using the new timetable, Monkseaton's Year 11 pupils recorded the best GCSE results in the school's thirty-nine-year history': 'School finds the secret of exam success: let teenagers have a lie-in', Hannah Devlin, *The Times*, 27/8/2011. http://www.thetimes.co.uk/tto/education/article3146950.ece

'In the late 1960s, the Stanford University psychologist Walter Mischel devised a simple test for three-year-olds at campus nursery': for a good account, see 'Don't: the secret of self-control', Jonah Lehrer, *New Yorker*, 18/5/2009. http://www.newyorker.com/reporting/2009/05/18/090518fa_fact_lehrer

'The research, led by Terrie Moffitt and Avshalom Caspi, of Duke University and King's College London, showed that children as young as three who had poor self-control skills were more likely to grow up with a wide range of health and social problems': 'A gradient of childhood self-control predicts health, wealth, and public safety', Moffitt, Caspi et al., *Proceedings of the National Academy of Sciences*, 24/1/2011. http://www.pnas.org/content/early/2011/01/20/1010076108

'A systematic review of thirty-four studies, led by Alex Piquero, of Florida State University, found in 2010 that role-playing exercises, rewarding self-control and several other interventions held promise for teaching delayed gratification. Another review, led by Adele Diamond, of the University of British Columbia, has found that computer training, martial arts and yoga can have a beneficial effect': 'Self-control interventions for children under age 10 for improving self-control and delinquency and problem behaviors', Piquero et al., *Campbell Systematic Reviews*, 2010:2. http://campbellcollaboration.org/lib/download/792. Adele Diamond and Kathleen Lee, 'Interventions Shown to Aid Executive Function Development in Children 4 to 12 Years Old', *Science*, 19/8/2011. http://www.sciencemag.org/content/333/6045/959

'Its potential to unlock the mechanisms by which the brain learns was highlighted in 2011 by a Royal Society working party chaired by Professor Uta Frith': 'Neuroscience: implications for education and lifelong learning', *Brain Waves* 2, 24/2/2011. http://royalsociety.org/policy/projects/brain-waves/education-lifelong-learning/

'The fine motor control needed to manipulate a pencil or pen develops fully only around the age of five – which may have important implications for teaching handwriting, which commonly begins at a younger age': Sarah-Jayne Blakemore and Uta Frith, *The Learning Brain*, Wiley-Blackwell, 2005.

'A seven-year study in Clackmannanshire, Scotland, suggested that children taught using phonics were three and a half years ahead of their peers in reading by the time they finished primary school': Johnston and Watson, *A Seven Year Study of the Effects of Synthetic Phonics Teaching on Reading and Spelling Attainment*, 2005. http://www.scotland.gov.uk/Publications/2005/02/20682/52383

'In 2006, an expert inquiry led by Sir Jim Rose, a former director of inspection at the Office for Standards in Education, found an "overwhelming case" for phonics': Rose review into the teaching of reading, March 2006. http://www.literacytrust.org.uk /assets/0000/1175/Rose_Review.pdf. For a detailed critique of the Rose review and the Clackmannanshire evidence, see 'Synthetic phonics and the teaching of reading: the debate surrounding England's "Rose Report" ', Dominic Wyse and Morag Styles, *Literacy*, Volume 41, Number 1, April 2007. http://www.ncne.co.uk/phdi/p1.nsf/ 19dd5ec20cdbd99a8025679d00581578/6f5c982e3b539c3d80257657005d8178/$FILE/ RoseEnquiryPhonicsPaperUKLA.pdf

'the government asked a team led by Carole Torgerson, then of the University of York, to examine the evidence for phonics': Torgerson et al., 'A systematic review of the research literature on the use of phonics in the teaching of reading and spelling', January 2006. https://www.education.gov.uk/publications/eOrderingDownload/ RB711.pdf

' "That was never done," Torgerson says': interview with author, 21/6/2011.

'When the Commons Science and Technology Committee challenged Carole Willis, chief scientific adviser at the Department for Children, Schools and Families': Science and Technology Committee, Evidence Check 1: Early Literacy Interventions, 18/12/2009. http://www.parliament.the-stationery-office.co.uk/pa/cm200910/cmselect/ cmsctech/44/44.pdf

'In June 2010, Michael Gove announced a radical plan to convert the 200 weakest primary schools in England into academies, free from local authority control, to improve standards': 'Michael Gove unveils plan to convert weakest primary schools into academies', Jessica Shepherd, *Guardian*, 16/6/2011. http://www.guardian. co.uk/politics/2011/jun/16/michael-gove-weakest-primary-schools-acadamies

' "When you do a carefully designed, well-conducted, well-reported trial of a policy, it leaves you with a result you can't ignore because there are no alternative explanations," says Torgerson': interview with author, 21/6/2011.

'Yet a National Audit Office report published in 2010 found that this target was not backed up by routine collection of official data so that ministers could be held to it': 'Educating the next generation of scientists', NAO report, 12/11/2010. http://www.nao .org.uk/publications/1011/young_scientists.aspx

'Carole Willis summed up the argument when she told the Science and Technology Committee': Science and Technology Committee, Evidence Check 1: Early Literacy Interventions, 18/12/2009, p. 19, para 51. http://www.parliament.the-stationery- office.co.uk /pa/cm 200910/cmselect/cmsctech/44/44.pdf

' "It's not unethical to do experiments in education," says Sir Mark Walport': interview with author; reported in 'Test teaching ideas before imposing them on children', Mark Henderson, *The Times*, 5/2/2007. http://www.thetimes.co.uk/tto/education/article1877959.ece

'Brain Gym, "an educational, movement based programme aimed at enabling children and adults to reach their potential", is a prime example': http://www.braingym.org.uk /about/BGTE%20Responses.pdf

'As Ben Goldacre explained in *Bad Science*': Goldacre, *Bad Science*, p. 14.

'In 2008, Sense About Science put together a briefing on Brain Gym': http://www.sense aboutscience.org/data/files/resources/55/braingym_final.pdf

'There was no evidence that removing impacted teeth in the absence of symptoms led to better outcomes – and the risks of infection and nerve damage far outweighed the benefits': 'Surgical removal of third molars', J. P. Shepherd, M. Brickley, *BMJ*, 1994, 309:620–21. http://www.bmj.com/content/309/6955/ 620.full

' "It is an extremely powerful and productive way to run a profession," Shepherd says.': interview with author, 14/1/2011.

'Alom Shaha, a physics teacher and film-maker, agrees': interview with author, 8/3/2011.

'He's also involved in an online journal club for science teachers': http://science. teachingjournalclub.org

' "Suppose we find that the more often people consult astrologers or psychics, the longer they live," she might begin. "Why might that be true?" ': Susan Blackmore, Edge.com, 2011 World Question Centre: 'What scientific concept would improve everybody's cognitive toolkit?' http://www.edge.org/q2011/q11_12.html #blackmore

'As Carl Sagan asked in *The Demon-Haunted World*: "If we teach only the findings and products of science" ': Sagan, *The Demon-Haunted World,* p. 21.

'It was only when an editor at *The Times* serendipitously asked me to cover the science brief that I really began to appreciate the rigour and beauty of science, to value it as the ultimate satisfier of a curious mind': the editor in question was Ben Preston. I remain very grateful to him.

'In 2006, a learning strand known as How Science Works (HSW) was introduced to the national curriculum': for further details, see Nuffield Foundation document http://www.nuffieldfoundation.org/how-science-works

'Its introduction, however, was controversial, with critics arguing that it has replaced hard facts with half-baked philosophical musing, that it was a symptom of dumbing-down': see for example 'Science elite rejects new GCSE as fit for the pub', Alexandra Blair and Mark Henderson, *The Times*, 11/10/2006. http://www.thetimes.co.uk /tto/news/uk/article1944010.ece

'Examination boards have often set questions that lack appropriate rigour; Alom Shaha describes some as "ambiguous, subjective and downright unscientific" ': interview with author, 8/3/2011.

'A 2009 GCSE biology question set by the AQA board invited candidates to match four "theories of how new species of plants and animals have developed" ': 'Creationism question in "misleading" science GCSE', Graeme Paton, *Daily Telegraph*, 6/7/2009. http://www.telegraph.co.uk/science/evolution/5749730 /Creationism-question-in-misleading-science-GCSE.html

'HSW has also suffered because the science teachers who are expected to teach it have not been given the professional training that is required if it is to enrich their lessons': see for example 'How science works isn't working in British schools', Alom Shaha, *New Scientist S-word* blog, 4/12/2009. http://www.newscientist.com/blogs /thesword/2009/12/how-science-works-isnt-working.html

' "How Science Works was a much needed change in the science curriculum," says Shaha': interview with author, 8/3/2011.

'as philosophers of science sometimes put it, most scientists know as little about the philosophy of science as fish know about water': see for example 'Is science losing its objectivity?', John Ziman, *Nature*, 29/8/1996. http://libweb.surrey.ac.uk/library/skills/ Science%20and%20Society/SS_1_Reading2.pdf

'As part of a drive to make the teaching profession "brazenly elitist", the Education Secretary decreed that only graduates with a lower second or better would henceforth be eligible for public funding to train as teachers': 'We need a brazenly elitist approach to education, say Tories', Francis Elliott, *The Times*, 19/1/2010. http://www.thetimes .co.uk/tto/news/politics/article2030656.ece

'One in four maths teachers in state secondary schools is not a specialist in the subject, and 16 per cent of lessons are not taught by maths graduates': Helen Moor et al., 'Mathematics and Science in Secondary Schools', DfE report, 2006. https:// www.education.gov.uk/publications/eOrderingDownload/RR708.pdf

'The Institute of Physics estimates that more than 4,000 specialist physics teachers are needed if every pupil is to have the opportunity to learn the subject from a specialist': http://www.iop.org/education/ltp/news/11/file_51731.pdf

'Gove's department has accepted this, and set an annual recruitment target of 925 physics graduates each year': 'Government announces specific physics teacher target', Institute of Physics, 4/2/2011. http://www.iop.org/news/11/feb/page_48715.html

'as only about 3,000 physics students graduate annually, the Department of Education is hoping to attract one in three of them': '4,000 shortfall in physics teachers', *Times Education Supplement*, 11/10/2010. http://newteachers.tes.co.uk/news/4000-shortfall- physics-teachers/45485

'One in four physics teachers who began training in 2008, one in five maths teachers and one in six chemistry teachers had a third': Alan Smithers and Pamela Robinson, *The Good Teacher Training Guide* 2010, p. 15. http://www.alansmithers .com/reports/ The_Good_Teacher_Training_Guide_2010.pdf

'even with generous new bursaries of up to £20,000 available for maths, physics and chemistry graduates who meet the 2:2 requirement': http://media.education.gov.uk /assets/files/pdf/t/training%20our%20next%20generation%20of%20outstanding% 20teachers.pdf

' "Teaching is of better quality where teachers hold qualifications in the subjects they teach," it declared in a 2010 report on science education': NAO, 'Educating the Next Generation of Scientists': http://www.nao.org.uk/publications/1011/young_scientists.aspx

'A government task force chaired by Sir Mark Walport, director of the Wellcome Trust, reached a similar conclusion': Science and Learning Expert Group report, 25/2/2010. http://interactive.bis.gov.uk/scienceandsociety/site/learning/2010/02/25/ new-science-and-learning-expert-group-report

' "I fear I'll be meeting more and more people who will tell me they hated physics at school – and it won't be their fault," says Alom Shaha': 'Where have all the passion- ate physics teachers gone?', Alom Shaha, *Guardian Science* blog, 13/12/2010. http://www.guardian.co.uk/science/blog/2010/dec/13/physics-teachers-shortage

'Some of the best evidence for this has emerged from a series of "state of the nation" assessments of science education by the Royal Society, the most recent of which, for the 16–19 age group, was published in February 2011': Andrew Hughes et al., 'Preparing for the transfer from school and college science and mathematics education to UK STEM higher education', Royal Society, 15/2/2011. http://royalsociety. org/uploadedFiles/Royal_Society_Content/education/policy/state-of-nation /2011_02_15-SR4-Fullreport.pdf

'the overall uptake of science at A-level and its equivalents is increasing': a good summary of recent increases is available from the Campaign for Science and Engineering at http://sciencecampaign.org.uk/?p=6997

'In the 2008/9 school year, about 299,000 teenagers in England and Wales were studying for at least one A-level or equivalent qualification': Hughes et al. http://royalsociety.org/uploadedFiles/Royal_Society_Content/education/ policy/state-of-nation/2011_02_15-SR4-Fullreport.pdf Table 3.1 p11

'Some 17 per cent of schools offering A-levels in England, 13 per cent in Wales and 43 per cent in Northern Ireland failed to enter a single candidate in physics, while a similar number entered just one or two': Hughes et al. See also Royal Society press release. http://royalsociety.org/news/Calls-for-A-level-reform-/

'A 2010 analysis of twenty-four countries by the Nuffield Foundation found that England, Wales and Northern Ireland were the only ones where fewer than one in five pupils studies maths in upper-secondary education': Nuffield Foundation, 'Is the UK an Outlier?', Jeremy Hodgen et al., 2010. http://www.nuffieldfoundation. org/sites/default/files/files/Is%20the%20UK%20an%20Outlier_Nuffield%20 Foundation_v_FINAL.pdf

'As a result, fewer than 10,000 British students graduate in physics, chemistry and maths from British universities each year': Royal Society report, 'The UK's science and mathematics teaching workforce', 10/12/2007, p. 98. http://royalsociety.org/ education/policy/state-of-nation/teaching-workforce

'This output, the Royal Society found, "leads to a deficit of STEM graduates available to enter employment in commerce and industry"': Royal Society press release, 15/2/2011. http://royalsociety.org/news/Calls-for-A-level-reform-/

'More than four in ten employers report difficulties recruiting graduates with appropriate STEM skills': Confederation of British Industry, 'Building for Growth', 9/5/2011. http://www.cbi.org.uk/pdf/20110509-building-for-growth.pdf

'To Dame Athene Donald, Professor of Physics at the University of Cambridge, who chairs the Royal Society's education committee, this "could be hugely damaging to the prospects of both the individual student and our nation as a whole"': Royal Society press release, 15/2/2011. http://royalsociety.org/news/Calls-for-A-level-reform-/

North of the border, about 50 per cent of all pupils taking Scottish Highers study a science subject': Hughes et al., Table 3.1, p. 11. http://royalsociety.org/ uploadedFiles /Royal_Society_Content/education/policy/state-of-nation/2011_02_15-SR4-Fullreport.pdf

'but the Department for Education admitted to the National Audit Office that such provision was "patchy"': NAO, 'Educating the Next Generation of Scientists': http://www.nao.org.uk/publications/1011/young_scientists.aspx

'Separate physics, chemistry and biology courses are more commonly available in schools in rich parts of the country, and in the independent sector, but the benefits aren't simply an artefact of this': NAO, 'Educating the Next Generation of Scientists' http://www.nao.org.uk/publications/1011/young_scientists.aspx

'In his introduction to the Royal Society's 2010 report into 5–14 education, Lord Rees of Ludlow, then its President, noted: "Children are innately curious about the natural world"': 'Primary Science and Mathematics Education: Getting the Basics Right', 2010. http://royalsociety.org/uploadedFiles/Royal_Society_Content /education/policy /state-of-nation/2010-07-07-SNR3-Basics.pdf

'There are 193,000 qualified teachers in England's 17,000 primary schools': 'Stat of the week – Number of primary school teachers', John Howson, *Times Education Supplement*, 29/5/2009. http://www.tes.co.uk/article.aspx?storycode= 6014558

'the Royal Society found that only 5,989 of them had a background in science and 3,903 had a degree in maths': 'Primary Science and Mathematics Education: Getting the Basics Right', 2010. http://royalsociety.org /uploadedFiles/Royal_Society_Content/education/policy/state-of-nation/2010-07-07-SNR3-Basics.pdf

'A 1997 study by Wynne Harlen and Colin Holroyd found that teachers short on confidence tend to use "safe" teaching methods that rely heavily on work books, underplay questioning and discussion, and avoid experiments that use unfamiliar equipment': 'Primary teachers' understanding of concepts of science: impact on confidence and teaching', Harlen/Holroyd, *Int J Sci Educ*, 19, 93–105.

'The Royal Society noted further that: "Teachers lacking a robust understanding of the subject matter and how to teach it are more likely to be influenced by the content of tests than those who have the confidence to know that effective teaching will achieve good results without teaching to the tests"': 'Primary Science and Mathematics Education: Getting the Basics Right', 2010. http://royalsociety.org /uploadedFiles/Royal_Society_Content/education/policy/state-of-nation/2010-07-07-SNR3-Basics.pdf

'The Royal Society predicts that the 2:2 requirement will "slash the supply of people into science and mathematics teaching by about 50 per cent"': 'Primary Science and Mathematics Education: Getting the Basics Right', 2010, p. 42. http://royalsociety.org/uploadedFiles/Royal_Society_Content/education/ policy/state-of-nation/2010-07-07-SNR3-Basics.pdf

'The Campaign for Science and Engineering (CaSE), and learned societies such as the Royal Society of Chemistry and the Institute of Physics, have joined the Royal Society in making this point to the minister': 'New teacher degree standard "will worsen shortage in maths and science"', Mark Henderson, *The Times*, 14/2/2011. http://www.thetimes.co.uk/tto/education/article2912629.ece

'"If you want to communicate science, if you really want to make people love and understand science, you can have a far more real and measurable impact as a science teacher," he says': interview with author, 8/3/2011.

'The foraging behaviour of the buff-tailed bumblebee, *Bombus terrestris*, is just the sort of material you'd expect to be examined in the pages of *Biology Letters*, a peer-reviewed journal published by the Royal Society': 'Blackawton bees', Blackawton PS et al., *Biology Letters*, 23/4/ 2011 7:168–172. http://rsbl.royalsocietypublishing.org /content/early/2010/12/18/rsbl.2010.1056.full Commentary by Laurence T. Maloney and Natalie Hempel de Ibarra http://rsbl.royalsocietypublishing.org/content /early/2010/12/09/rsbl.2010.1057

'As Strudwick put it: "Science shouldn't be seen as something that is detached from the real world – it's just a certain way of looking at things. This project represents a completely different way of working and learning that I hope will be taken up by other schools and in other subjects"': 'Bee plus? The buzz created by these pupils is better than an A*', Mark Henderson, *The Times*, 22/12/2010. http://www.thetimes .co.uk/tto/science/biology/article2852120.ece

'At the Simon Langton Grammar School in Canterbury, sixth-form students are every year given the opportunity to take part in real scientific research – a recent project involved helping to clone a gene associated with multiple sclerosis': 'Researchers in residence': http://wellcometrust.wordpress .com/2010/11/26/real-school-real-science

'Research Councils UK and the Wellcome Trust run a Researchers in Residence scheme': http://www.researchersinresidence.ac.uk/cms

'David Spiegelhalter, the Winton Professor of the Public Understanding of Risk at the University of Cambridge, has developed a fantastic range of aids for teaching schoolchildren about risk and statistics': http://understanding uncertainty.org

'Oxford's Centre for Evidence Based Medicine provides outstanding resources for adding the subject to science lessons': http://www.cebm.net/index.aspx?o=1083

'Then there's *I'm a Scientist, Get me out of here!*': http://imascientist.org.uk

7 Doing Science Justice

'On the afternoon of 24 January 2008, David Chenery-Wickens walked into a London police station': for details of the case, see 'Forensic Science Service Case Studies: Diane Chenery-Wickens', http://www.forensic.gov.uk/html/media/case-studies/ f-57.html; 'David Chenery-Wickens: railway enthusiast and wife murderer', Helen Nugent, *The Times*, 3/3/2009. http://www.thetimes.co.uk/tto/news/uk/crime/article 1875772.ece. A good account of the Chenery-Wickens case is available at 'Britain's forensic scientists are the best in the world: So why is their elite force being disbanded?', Adam Luck, *Daily Mail*, 20/4/2011. http://www.dailymail.co.uk /home/moslive/article-1377202/Britains-forensic-scientists-best-world-So-elite-force-disbanded.html

' "There is no justification for the uncertainty and costs of trying to restructure and retain the business," the Home Office declared': Home Office statement, 14/12/2010. http://www.homeoffice.gov.uk/publications/about-us/parliamentary-business/written-ministerial-statement/forensic-science-wms

' "Purely commercial suppliers in such a competitive structure are forced to reduce costs to levels that cannot support the type of research, innovation and attention to case-specific needs that has characterised the commitment to service shown by the FSS," noted an international group of experts in a letter to *The Times*': 28/12/2010. http://www.thetimes.co.uk/tto/opinion/letters/article 2855879.ece

'The conclusions of a House of Commons Science and Technology Committee inquiry were equally damning': Science and Technology Committee report, 'The Forensic Science Service', 1/7/2011. http://www.publications.parliament.uk/pa/cm201012/ cmselect/cmsctech/855/855.pdf

'The closure had also been rushed, so that police authorities were forced to sign contracts with private forensic services without having time for "due diligence" on their credentials': 'Forensic Science Service closure forces police to use untested private firms', http://www.guardian.co.uk/uk/2011/aug/03/forensic-science-service-closure-police

'"We do need to ask a very serious question whether the current proposals carry the risk of damaging the system of justice," said Andrew Miller, the committee's chairman': 'Speed of forensic science cuts "harms the quality of evidence"', Mark Henderson, *The Times*, 1/7/2011. http://www.thetimes.co.uk/tto/news/article 3080718.ece

'In 1996, Sally Clark, a solicitor from Cheshire, gave birth to her first child, Christopher': details of Sally Clark's trial are from the Royal Statistical Society report into the case, available at http://www.sallyclark.org.uk/RSS.html

'In January 2006, Mary Jackson (not her real name) was attacked in a parking lot in Sacramento, California, and forced to perform oral sex on her assailant': 'Fallible DNA evidence can mean prison or freedom', Linda Geddes, *New Scientist*, 11/8/2010. http://www.newscientist.com/article/mg20727733.500-fallible-dna-evidence-can-mean-prison-or-freedom.html?page=1 http://www.newscientist.com/article/mg20727743.300-how-dna-evidence-creates-victims-of-chance.html?page=1

'Working with Itiel Dror, of University College London, and Greg Hampikian, of Boise State University in Idaho, Geddes obtained DNA samples from a gang rape in Georgia, of which Kerry Robinson was convicted': I.E. Dror, G. Hampikian, 'Subjectivity and bias in forensic DNA mixture interpretation', *Sci. Justice* (2011) doi:10.1016/j.scijus.2011.08.004

'Geddes's findings particularly question the wisdom of allowing police forces to commission forensic analysis from in-house labs, which are usually unaccredited and lack the arm's-length independence of providers like the condemned FSS': 'Forensic Science Service closure forces police to use untested private firms', Alan Travis, *Guardian*, 3/8/2011. http://www.guardian.co.uk/uk/2011/aug/03/forensic-science-service-closure-police

'"There's pressure from prosecutors and investigators to get a result," Geddes says': interview with author, 20/9/2011.

'In 2009, it decided it might be able to use science to distinguish genuine asylum-seekers fleeing war or persecution from economic migrants falsely claiming asylum to get into the country': 'Scientists Decry "Flawed" and "Horrifying" Nationality Tests', John Travis, *Science Insider*, 29/9/2009. http://news.sciencemag.org/scienceinsider/2009/09/border-agencys.html

'The project was eventually scrapped in 2011, wasting more than £100,000 of public money': 'DNA test for bogus refugees scrapped as expensive flop', Rebecca Hill and Mark Henderson, *The Times*, 17/6/2011. http://www.thetimes.co.uk/tto/news/uk/article3064981.ece

'The Department for Work and Pensions (DWP) introduced a pilot based on similarly questionable science in 2009, this time of "voice risk analysis" software designed to assess whether benefit claimants were lying': 'The truth is on the line', Charles Arthur, *Guardian*, 12/3/2009. http://www.guardian.co.uk/technology/2009/mar/12/voice-analysis-system-vra

'and the technology had been criticized in the scientific literature as "at the astrology end of the validity spectrum"': Eriksson and Lacerda, *International Journal of Speech, Language and the Law*, 2007. http://www.scribd.com/doc/9673590 /Eriksson-Lacerda-2007

'The tests failed: of the twenty pilots, only five reported a correlation between claims judged as high risk and actual fraud': 'The Application of Voice Risk Analysis within the Benefits System', DWP Evaluation Report, September 2010. http://www.dwp.gov.uk/docs/vra-evaluation.pdf

'As Francisco Lacerda, Professor of Phonetics at Stockholm University, put it: "I would be surprised if they had reached another conclusion"': 'Government abandons lie detector tests for catching benefit cheats', Ian Sample, *Guardian*, 9/11/2010. http://www.guardian.co.uk/science/2010/nov/09/lie-detector-tests-benefit-cheats

'In 2008, the Iraqi Interior Ministry signed a £19 million contract with a British company called ATSC to supply 800 bomb detection devices': 'Iraq Swears by Bomb Detector U.S. Sees as Useless', Rod Nordland, *New York Times*, 3/11/2009. http://www.nytimes.com/2009/11/04/world/middleeast/04sensors. html

'The British government eventually banned the export of the devices in January 2010, after ordering tests that showed "the technology used in the ADE651 and similar devices is not suitable for bomb detection"': BBC *Newsnight*, Caroline Hawley and Meirion Jones, 22/1/2010. http://news.bbc.co.uk/1/hi/programmes/newsnight /8471187.stm

'It was lunchtime of Friday, 30 October 2009, and Professor David Nutt was preparing his slides for an academic meeting on addiction when his mobile rang': interview with author, 13/7/2011. See also David Nutt's submission to the Commons Science and Technology Committee. http://www.parliament.uk/documents/upload/091118-a-letter-from-prof-nutt.pdf

'The reason for his dismissal was a comment piece Nutt had written for the *Guardian* the previous day': 'The cannabis conundrum', David Nutt, *Guardian*, 29/10/2009. http://www.guardian.co.uk/commentisfree/2009/oct/29/cannabis-david-nutt-drug-classification

'But to Johnson, Nutt had "crossed a line" by "campaigning against government decisions": interview with Sky News, 1/11/2009. http://news.sky.com /skynews/Home/UK-News/Drugs-Row-Second-Member-Of-Governments-Advisory-Council-On-Drugs-Quits/Article/200911115428280

'"It is important that the Government's messages on drugs are clear and as an adviser you do nothing to undermine the public understanding of them," the Home Secretary wrote in Nutt's letter of dismissal. "I cannot have public confusion between scientific advice and policy"': Press Association report, 30/10/2009. http://www.independent.co.uk/news/uk/politics/drug-adviser-sacked-over-lsd-claims-1812091.html

'As the council was still deliberating, Jacqui Smith's spokesperson told the BBC: "The Government firmly believes that ecstasy should remain a class A drug"': 'Ecstasy: Class A drug?', Mark Easton, BBC, 26/9/2008. http://www.bbc.co.uk/blogs/thereporters/markeaston/2008/09/what_is_the_point_of.html

'In January 2009, Nutt published a paper in the *Journal of Psychopharmacology* that compared the risks of two popular activities': 'Equasy – an overlooked addiction with implications for the current debate on drug harms', D. J. Nutt, *J Psychopharmacol.*, 2009 Jan; 23(1):3–5. http://www.ncbi.nlm.nih.gov/pubmed/19158127

'As about thirty deaths a year involve use of ecstasy, the risks involved are broadly comparable': http://www.patient.co.uk/doctor/Ecstasy-%28MDMA% 29.htm

'"I had the Home Secretary ringing me up and remonstrating with me about the reprehensible thing I'd done," Nutt recalls': interview with author, 13/7/2011.

'So routine did he consider the dismissal that he consulted neither Sir John Beddington, the government's chief scientist, nor Lord Drayson, the Science Minister': 'Government backs down on science freedom demands', Mark Henderson, *The Times*, 6/11/2009. http://www.thetimes.co.uk/tto/news/politics/article2030101.ece

'"It is critical that Chairs and members of independent scientific committees are not just independent of Government but are positively encouraged to provide the best possible interpretation of the available scientific data, whether or not that interpretation fits with the current political view," said Professor Chris Higgins': Science Media Centre email, 2/11/2009. http://www.sciencemediacentre.org/pages/press_releases/09-11-02_david_nutt.htm

'Lord Drayson's consternation at the sacking was revealed when emails he sent to the Prime Minister's office were leaked': 'Cabinet in drug war over sacking', David Wooding, *Sun*, 3/11/2009. http://www.thesun.co.uk/sol/homepage/news/2711166/Cabinet-in-drug-war-over-sacking-of-Professor-David-Nutt.html

'Sir John Beddington agreed publicly with Nutt's opinion that alcohol and tobacco are more dangerous than cannabis': 'Science chief backs cannabis view', Pallab Ghosh, BBC, 3/11/2009. http://news.bbc.co.uk/1/hi/8340318.stm

'Drayson and Beddington began to share their fears that experts might begin to refuse invitations to sit on advisory panels unless bridges were mended': 'Government backs down on science freedom demands', Mark Henderson, *The Times*, 6/11/2009. http://www.thetimes.co.uk/tto/news/politics/article 2030101.ece

'Sense About Science organized a statement of principles for the treatment of expert advice'. http://www.senseaboutscience.org.uk/index.php/site/project/ 421

'It accepted most of the Sense About Science proposals, including, importantly, the right of committees and their chairs to communicate their findings to the public even when they are rejected': 'Principles of Scientific Advice to Government': http://www.bis.gov.uk/go-science/principles-of-scientific-advice-to-government

'Any adviser who speaks out of turn can be accused of undermining trust and dealt with accordingly': for fuller criticism, see Sense About Science's response. http://www.senseaboutscience.org.uk/index.php/site/project/471

'In December 2010, the new Conservative–Liberal Democrat Government published a Police Reform and Social Responsibility Bill, with a clause that sought to address this anomaly': 'Ministers to set drug policy without scientists', Mark Henderson, *The Times*, 6/12/2010. http://www.thetimes.co.uk/tto/science/medicine/article2833559.ece

'Raabe had previously distinguished himself as the author of a report to the Canadian Parliament that stated': document now withdrawn from the web; quoted in 'GP's new role reopens controversy over government drugs advice', Mark Henderson, *The Times*, 20/1/2011. http://www.thetimes.co.uk/tto/health/news/article2882755.ece

'Evan Harris, the former MP and champion of science, tipped off *The Times* and the BBC, both of which highlighted Raabe's opinions': 'GP's new role reopens controversy over government drugs advice', Mark Henderson, *The Times*, 20/1/2011. http://www.thetimes.co.uk/tto/health/news/article2882755.ece; 'Another ACMD member threatens to quit', Mark Easton, BBC, 20/1/2011. http://www.bbc.co.uk/blogs/thereporters/markeaston/2011/01/another_acmd_member_threatens.html

'Skeptical bloggers such as Tom Chivers and Andy Lewis pored over Raabe's published views, and tore into the dubious evidence with which he supports them': Chivers: 'Dr Hans-Christian Raabe of the ACMD: it doesn't matter that you're a Christian, you're just no good with evidence', Tom Chivers, *Daily Telegraph*, 7/2/2011.   http://blogs.telegraph.co.uk/news/tomchiversscience/100075023/dr-hans-christian-raabe-of-the-acmd-it-doesnt-matter-that-youre-a-christian-youre-just-no-good-with-evidence; Lewis: 'When the Regulator Believes in Fairies, Who Protects the Public?', Andy Lewis, *The Quackometer*, 22/1/2011. http://www.quackometer.net/blog/2011/01/when-the-regulator-believes-in-fairies-who-protects-the-public.html

'Just two weeks after his appointment, Raabe was sacked, for failing to declare his authorship of the paedophilia paper, which "raises concerns over his credibility to provide balanced advice on drug misuse issues"': 'I'm victim of PC vendetta, says Christian drug expert as he is sacked from Government advisory panel before he even started', James Slack and Ian Drury, *Daily Mail*, 7/2/2011. http://www.dailymail.co.uk/news/article-1354325/Christian-drug-expert-Hans-Christian-Raabe-sacked-Government-advisory-panel.html#ixzz1ayxJNMuF

'Randomized trials show that prescribing heroin to long-term addicts significantly improves both health and social outcomes': 'Supervised injectable heroin or injectable methadone versus optimised oral methadone as treatment for chronic heroin addicts in England after persistent failure in orthodox treatment (RIOTT): a randomised trial', Strang et al., *Lancet*, 2010. http://www.thelancet.com/journals/lancet/article/PIIS0140-6736% 2810%2960349-2/abstract; heroin prescription: 'Cost utility analysis of co-prescribed heroin compared with methadone maintenance treatment in heroin addicts in two randomised trials', Marcel G. W. Dijkgraaf et al., *BMJ*, 2005. http://www.bmj.com/content / 330/7503/1297

'There is good evidence that exchange programmes significantly reduce sharing of needles, and some evidence that this improves HIV infection rates': 'Evidence for the effectiveness of sterile injecting equipment provision in preventing hepatitis C and human immunodeficiency virus transmission among injecting drug users: a review of reviews', Norah Palmateer, *Addiction*, 2010. http://onlinelibrary.wiley.com /doi/10.1111/j.1360-0443.2009.02888.x/abstract

'they do not encourage people to begin using heroin, but rather allow existing users to manage their addiction more safely': http://www.ncbi.nlm.nih.gov/pubmed/12794555 J Acquir Immune Defic Syndr. 2003 Jun 1;33(2):199-205. Needle exchange and injection drug use frequency: a randomized clinical trial. Fisher DG, Fenaughty AM, Cagle HH, Wells RS.

'National Academy of Sciences recommended as long ago as 1995 that "well-implemented needle exchange programmes can be effective in preventing the spread of HIV and do not increase the use of illegal drugs," and the Department of Health and Human Services reached a similar conclusion three years later': 'Preventing HIV Transmission: The Role of Sterile Needles and Bleach', Institute of Medicine, 1995. http://www.nap.edu/openbook.php?isbn=0309052963

'"There was a ban on using federal money for needle-exchange programmes unless we could show that they reduce the spread of infectious disease and do not increase drug use," he explains': Alan Leshner, interview with author, 18/2/2011.

'As Chris Mooney recounts in *The Republican War on Science*': Mooney, *The Republican War on Science*, pp. 240–41.

'the statement by Gordon Brown, as Prime Minister, that some powerful strains of skunk are "lethal" is factually incorrect': Brown is reported as saying this by Reuters, 30/4/2008. http://uk.reuters.com/article/2008/04/30/uk-britain-cannabis-for-uk-pack-only-idUKL2973937220080430?pageNumber=1&virtualBrandChannel=0

'Research led by Matthew Hickman, of the University of Bristol, shows that it would be necessary for at least 2,800 cannabis users, and perhaps as many as 10,870, to stop smoking the drug to prevent a single case of psychosis': Hickman et al., *Addiction*, November 2009, 104(11):1856–61. 'If cannabis caused schizophrenia – how many cannabis users may need to be prevented in order to prevent one case of schizophrenia?' England and Wales calculations. http://www.ncbi.nlm.nih.gov/pubmed/19832786

'Almost exactly a year after his sacking, David Nutt published a paper in the *Lancet* that assessed the harms caused by twenty drugs according to sixteen criteria, such as addictive potential, known health risks and social impact': 'Drug harms in the UK: a multicriteria decision analysis', Nutt, King and Phillips, *Lancet*, Volume 376, Issue 9752, Pages 1558–65, 6/11/2010. http://www.thelancet.com/journals/lancet/article/PIIS0140-6736%2810% 2961462-6/abstract A good lay summary of this paper is available on David Nutt's blog. http://profdavidnutt.wordpress.com/2010/12/09/drug-harms-paper-a-summary

'"It's true that it's different, but that's like saying wine is different from beer," Nutt says': interview with author, 13/7/2011.

'The British Crime Survey showed no change in patterns of overall cannabis use, and a trend for slightly lower use among young people that began in 1998 continued. The police estimated that the new regime saved 199,000 hours of officers' time in its first year': 'Drug classification: making a hash of it', Commons Science and Technology Committee report, 31/7/2006, para 48. http://www.publications.parliament.uk/pa/cm200506/cmselect/cmsctech/1031/1031.pdf

'John Strang, director of the National Addiction Centre, gave an elegant summary of the gateway theory to the Commons Science and Technology Committee when it prepared its 2006 report, *Drugs Policy: Making a Hash of It*': 'Drug classification: making a hash of it', Commons Science and Technology Committee report, 31/7/2006, para51. http://www.publications.parliament.uk/pa/cm200506/cmselect/cmsctech/1031/1031.pdf

'When the ACMD considered the issue in 2002, it concluded: "Even if the gateway theory is correct, it cannot be a very wide gate as the majority of cannabis users never move on to Class A drugs". RAND Europe, which evaluated the evidence for the Commons committee, found that "the gateway theory has little evidence to support it despite copious research" ': 'Drug classification: making a hash of it', Commons Science and Technology Committee report, 31/7/2006, para 53. http://www. publications.parliament.uk/pa/cm200506/cmselect /cmsctech/1031/1031.pdf

' "There is no evidence that classification makes any difference at all," says Nutt. "It has never been studied. It's an act of faith. This should be a fundamental question: do these laws do what they're supposed to do? But we don't ask it" ': interview with author, 13/7/2011.

'These pilots were deemed a success, and DTTOs were rolled out nationwide in 2000': Pilot evaluation available at http://webarchive.nationalarchives.gov.uk /20110218135832/ http://rds.homeoffice.gov.uk/rds/pdfs/hors212.pdf. DTTOs were phased out and replaced by similar Drug Rehabilitation Requirements in 2005.

'When the pilot programme was started, Sheila Bird, of the MRC Biostatistics Unit, immediately sounded alarm bells': interview with author, 24/1/2011.

'In 2002, it asked Bird, with Gillian Raab and Helen Storkey, of Napier University in Edinburgh, to advise on the design of a pilot of Project Blueprint': Raab, Storkey, Bird, October 2002. http://www.mrc-bsu.cam.ac.uk/Publications/PDFs/SampleSize Calculations2002.pdf

'The evaluation document was published in 2009, and guess what? The samples, it concluded, were too small to generate significant data': 'Blueprint Drugs Education: The Response of Pupils and Parents to the Programme', Executive Summary, autumn 2009. http://www.drugeducationforum.com/images/dynamicImages/7327_359646.pdf

'As Mark Easton, the BBC's Home Affairs Editor, wrote when he revealed the results': 'Project Blueprint: "Not sufficiently robust" ', Mark Easton, BBC, 17/9/2009. http://www.bbc.co.uk/blogs/thereporters/markeaston/2009/09/project_blueprint.ht-ml?page=8

'Easton established that the true picture was worse even than Scholar realized': 'Knife stats row: The plot thickens', Mark Easton, BBC, 26/2/2009. http://www.bbc.co.uk /blogs/thereporters/markeaston/2009/02

'Under this initiative, which is being piloted at Peterborough Prison, a company called Social Finance is putting its own money into a mentoring programme': http://www.socialfinance.org.uk/sites/default/files/SF_Peterborough_SIB.pdf

'described by Michael Green and Matthew Bishop, the authors of *The Road From Ruin: A New Capitalism for a Big Society*, as "the best Big Society proposal" ': 'Save the Big Society, send for big business', Matthew Bishop and Michael Green, *The Times*, 14/2/2011. http://www.thetimes.co.uk/tto/opinion/columnists/article2912888 .ece

'Yet as Bird points out, it also has an intrinsic weakness': interview with author, 24/1/2011.

'it was specifically chosen as the study site, and the characteristics that caused it to be chosen could plausibly account for the programme's success or failure': 'Lessons learned from the planning and early implementation of the Social Impact Bond at

HMP Peterborough', Emma Disley et al., Ministry of Justice, May 2011. http://www.justice.gov.uk/downloads/publications/research-and-analysis/moj-research/social-impact-bond-hmp-peterborough.pdf

' "All these problems could have been averted when thinking through how to design these pilots," says Bird': interview with author, 24/1/2011.

'As a maxillofacial surgeon, Professor Jonathan Shepherd's clinical practice often involves reconstructive work following severe facial injuries': interview with author, 14/1/11.

'The results, published in the *British Medical Journal* in June 2011': Florence, Shepherd, Brennan and Simon, 'Effectiveness of anonymised information sharing and use in health service, police, and local government partnership for preventing violence related injury: experimental study and time series analysis', *BMJ* 2011, 342:d3313. http://www.bmj.com/content/342/bmj.d3313

'Then Shepherd designed an RCT': Alison L. Warburton and Jonathan P. Shepherd, 'Effectiveness of toughened glassware in terms of reducing injury in bars: a randomised controlled trial', *Injury Prevention*, 2000; 6; 36–40. http://www.ihra.net/files/2011/07/21/09.3_Warburton_-_Effectiveness_ of_Toughened_Glassware_.pdf

' "We replaced the entire pint pot stock of fifty-seven licensed premises in the West Midlands and South Wales with tempered or non-tempered glass," Shepherd recalls': interview with author, 14/1/11.

'Chris Mullin, the former Labour MP, recalls in his diaries the contribution of a young Conservative backbencher to a Home Affairs Select Committee inquiry that recommended a harm-reduction approach to cannabis control': Chris Mullin, *A View from the Foothills* (Profile Books, 2009).

8 Placebo Politics

' "Alternative remedies which could be dangerous are being targeted by the Government in a major drive to improve health and welfare," read the departmental press release': 'Alternative pet remedies: Government clampdown', 17/12/2010. http://www.defra.gov.uk/news/2010/12/17/pet-remedies/

'Homeopathy is a system of medicine first devised in the 1790s by a German doctor named Samuel Hahnemann': for a good account of the history of homeopathy (and for much more information on the evidence base for alternative medicine) see Edzard Ernst and Simon Singh, *Trick or Treatment: Alternative Medicine on Trial*, Bantam Press, 2008.

'At 12C, the remedy becomes so dilute that it is unlikely to contain even a single molecule of the original active ingredient: it is just water': 'Homeopathic dilutions', John Jackson, *UK Skeptics*. http://www.ukskeptics.com/homeopathic-dilutions.php

'To ingest one molecule of the active ingredient at this dilution, it would be necessary to consume 10^{41} pills – a trillion times the volume of the Earth': the necessary calculations are well explained at http://sciencebasedparenting.com/2010/11/18/make-your-own-homeopathic-remedies-and-play-dough

'To get a single molecule of that, you'd have to ingest 10^{400} molecules of solvent. There are only about 10^{80} atoms in the observable universe': see Centre for Inquiry, *Okanagan* blog. http://www.cfiokanagan.org/published-articles/os3-balancing-science-and-homeopathy

'At such dilutions, it might not surprise you to learn that only a single duck need be slaughtered to keep the world in oscillococcinum for a year': 'Flu Symptoms? Try Duck' , Dan McGraw, *US News & World Report*, 9/2/1997. http://www.usnews. com/usnews/biztech/articles/970217/archive_006221.htm

' "Molecules are very small," explains David Colquhoun, Professor of Pharmacology at University College London': email to author, 24/9/2011.

'As Colquhoun told the Commons Science and Technology Committee when it reviewed the evidence for homeopathy in 2010: "If homeopathy worked, the whole of chemistry and physics would have to be overturned" ': Science and Technology Committee Evidence Check 2: Homeopathy, 22/2/2010, para 59. http://www.publications. parliament.uk/pa/cm200910/cmselect/cmsctech/45/45.pdf

'The Cochrane Collaboration, a clearing house for systematic reviews, has found no evidence to support homeopathy as a treatment for asthma, dementia, attention deficit/hyperactivity disorder or induction of labour, or for palliative care in cancer': Cochrane on asthma: McCarney, R. W., Linde, K., Lasserson, T. J., 'Homeopathy for chronic asthma', Cochrane Database of Systematic Reviews 1999, Issue 1, Art. No.:CD000353. DOI: 10.1002/14651858.CD000353.pub2. http://www2.cochrane. org/reviews/en/ab000353.html. Dementia: This record should be cited as: McCarney R. W., Warner, J., Fisher, P., van Haselen, R., 'Homeopathy for dementia.', Cochrane Database of Systematic Reviews 2003, Issue 1, Art. No.:CD003803. DOI: 10.1002/14651858.CD003803. http://www2.cochrane.org/reviews/en/ab003803.html ADHD: Heirs, M., Dean, M. E., 'Homeopathy for attention deficit/hyperactivity disorder or hyperkinetic disorder.', Cochrane Database of Systematic Reviews 2007, Issue 4, Art. No.: CD005648. DOI: 10.1002/14651858.CD005648.pub2 http://www2. cochrane.org/reviews/en /ab005648.html. Cancer: http://www2.cochrane.org/ reviews /en/ab004845.html. Kassab, S., Cummings, M., Berkovitz, S., van Haselen, R., Fisher, P., 'Homeopathic medicines for adverse effects of cancer treatments', Cochrane Database of Systematic Reviews 2009, Issue 2, Art. No.: CD004845. DOI: 10.1002/14651858.CD004845.pub2 Induction: Smith, C. A., 'Homoeopathy for induction of labour.', Cochrane Database of Systematic Reviews 2003, Issue 4, Art. No.:CD003399. DOI: 10.1002/14651858.CD003399. http://www2.cochrane. org/reviews/en/ab003399.html The *Stuff and Nonsense* blog summarizes all this evidence superbly. http://jdc325.wordpress.com/2010/02/05/cochrane-reviews-of-homeopathy

'Edzard Ernst, Professor of Complementary Medicine at the University of Exeter, has shown that the more rigorous the methodology of the trial, the less likely it is to suggest that homeopathy works': E. Ernst, 'A systematic review of systematic reviews of homeopathy', *Br J Clin Pharmacol.*, December 2002; 54(6): 577–582. http://www.ncbi.nlm.nih.gov/pmc/articles/PMC1874503/?tool=pmcentrezdoi: 10.1046/j.1365-2125.2002.01699.x

'When the Commons Science and Technology Committee asked Mike O'Brien, the Labour health minister, whether he had any evidence that homeopathic remedies were better than placebos, he replied: "The straight answer is no" ': Science and Technology Committee Evidence Check 2: Homeopathy, 22/2/2010. http://www. publications.parliament.uk/pa/cm200910/cmselect/cmsctech/45/45.pdf paragraph 70

'Sir John Beddington, the government's chief scientist, was still more explicit. "I have made it completely clear that there is no scientific basis for homeopathy beyond the placebo effect and that there are serious concerns about its efficacy," he told the committee': 'NHS funding for homeopathy risks misleading patients, says chief scientist', Ian Sample, *Guardian*, 27/10/2010. http://www.guardian.co.uk/science /2010/oct/27/nhs-funding-homeopathy-chief-scientist

' "We should not take the view that patients should not be able to have homeopathic medicine when they want it," O'Brien told the committee': Science and Technology Committee Evidence Check 2: Homeopathy, 22/2/2010. http://www.publications. parliament.uk/pa/cm200910/cmselect/cmsctech/45/45.pdf Paragraph Ev74, question 248

' "The overriding reason for NHS provision is that homeopathy is available to provide patient choice," the government said': Government response to select committee report, p. 5, para 9. http://www.dh.gov.uk/prod_consum_dh/groups/dh_ digitalassets /@dh/@en/@ps/documents/digitalasset/dh_117811.pdf

'While the government does not audit such spending, the Society of Homeopaths estimates NHS funding at about £4 million a year': British Homeopathic Society website. http://www.britishhomeopathic.org/media_centre/facts_about_homeopathy/ nhs_referrals.html

'the Royal London Homeopathic Hospital was refurbished recently at a cost of £20 million': Science and Technology Committee Evidence Check 2: Homeopathy, 22/2/2010. http://www.publications.parliament.uk/pa/cm200910/cmselect/cmsctech/ 45/45.pdf p6, paragraph 14

'In the US, the National Center for Complementary and Alternative Medicine (NCCAM) has an annual budget of about $130 million'. http://nccam.nih.gov /about/budget /congressional

' "Whether it's a medicine you buy, or one prescribed for you as part of a course of treatment, it's reassuring to know that all medicines available in the UK are subject to rigorous scrutiny by the MHRA before they can be used by patients," the agency's mission statement declares. "This ensures that medicines meet acceptable standards on safety, quality and efficacy" ': MHRA mission statement. http://www.mhra.gov.uk /Aboutus/Whatweregulate/index.htm

'In 2006, the MHRA changed its rules so that suppliers of homeopathic remedies could make medical claims without any supporting scientific data': Science and Technology Committee Evidence Check 2: Homeopathy, 22/2/2010.http: //www.publications.parliament.uk/pa/cm200910/cmselect/cmsctech/45/45.pdf p30 chapter3

'Michael Baum, Professor Emeritus of Surgery at UCL, put it another way: "This is like licensing a witches' brew as medicine so long as the bat wings are sterile": 'Doctors attack natural remedy claims', Nigel Hawkes and Mark Henderson, *The Times*, 1/9/2006. http://www.thetimes.co.uk/tto/news/uk/article 1942486.ece

' "If regulation was applied to homeopathic medicines as understood in the context of conventional pharmaceutical medicines, these products would have to be withdrawn from the market as medicines," it said': Government response to select committee report, p. 5, para 9. http://www.dh.gov.uk/prod_consum_dh/groups/dh_digitalassets /@dh/@en/@ps/documents/digitalasset/ dh_117811.pdf

'Peter Hain, a Cabinet minister under Labour, thinks that homeopathy cured his son's eczema': speech to The Prince of Wales Foundation for Integrated Health, 12/10/2005. http://www.peterhain.org/default.asp?pageid=194&mpageid=89&groupid=2. Hain said: 'So instead, we turned to complementary medicine. And, with the help of home-opathy and tight restrictions on the sort of food that our son could eat – avoiding in particular gluten and milk products – both ailments went away'. This proves another point about alternative medicine: often it's used in combination with something that might actually work – dietary changes in this case – and gets an unwarranted slice of the credit.

'As Northern Ireland Secretary in 2007, he launched a £200,000 project to allow GPs in Belfast and Londonderry to refer more patients for homeopathy and acupuncture': 'Peter Hain and GetwellUK: pseudoscience and privatisation in Northern Ireland', David Colquhoun, DC's *Improbable Science* blog, 9/2/2007. http://www.dcscience. net/?p=33

'Anne Milton, the Tory Health Minister who responded to the select committee report, believes her experience as a nurse – and her grandmother's experience as a homeopathic nurse – taught her that homeopathy is effective': http://skeptical-voter.org /wiki/index.php?title=Anne_Milton#Homeopathy

'Senator Tom Harkin, who secured funding for it because he thinks his allergies were cured by an alternative therapist who was later forced to stop making false claims for his bee-pollen supplements by the Federal Trade Commission': 'Cures or "Quackery"? How Senator Harkin shaped federal research on alternative medicine', Stephen Budiansky, *US News & World Report*, posted 7/9/95. http://www .usnews.com/usnews/culture/articles/950717/archive_032434.htm

'In 2009, [Harkin] betrayed his abject misunderstanding of how science works': 'Senator Tom Harkin: NCCAM has "fallen short" because it hasn't validated enough woo', *Respectful Insolence* blog, 2/3/2009. http://scienceblogs.com/insolence/ 2009/03/maybe_nccam_isnt_so_bad_after_allnahhh.php

'The MHRA has acknowledged receiving at least seven letters from him about the regulation of alternative medicines': 'HRH "meddling in politics"', David Colquhoun, DC's *Improbable Science* blog, 12/3/2007. http://www.dcscience .net/?p=89

'The Prince reportedly buttonholed Andy Burnham, then the Health Secretary, on the subject at a reception in 2009, and recently met Andrew Lansley, Burnham's Conservative successor, to discuss it': Burnham: 'Prince Charles urges government to protect alternative medicine', Richard Alleyne, *Daily Telegraph*, 1/12/2009. http://www.telegraph.co.uk/health/healthnews/6701717/Prince-Charles-urges-govern-ment-to-protect-alternative-medicine.html. Lansley: 'H.R.H. The Prince Minister: Charles accused of meddling after he summons seven senior Ministers to Clarence House in just ten months', Chris Hastings and Simon Walters, *Mail on Sunday*, 3/7/2011. http://www.dailymail.co.uk/news/article-2010673/Prince-Charles-accused-meddling-summoning-7-senior-Ministers-Clarence-House-10-months.html

'It persuaded the Department of Health to give it £900,000 to produce a patient guide to alternative therapies that contained "numerous misleading and inaccurate claims concerning the supposed benefits", according to Edzard Ernst': Prince's guides to alternative medicine are "misleading and inaccurate", Mark Henderson, *The Times*, 17/4/2008; see also letter to *The Times* of same date from Edzard Ernst and Simon Singh.

'In 2005, it commissioned the Smallwood Report into alternative medicine, which argued that wider NHS provision would save money. Professor Ernst described its conclusions as "outrageous and deeply flawed" ': see 'Prince plots alternative treatments for the NHS', Mark Henderson and Andrew Pierce, *The Times*, 24/8/2005. http://www.thetimes.co.uk/tto/news/uk/article1937237.ece

'Sir Michael Peat, the Prince's private secretary and the Foundation's acting chairman, made an official complaint to Exeter's Vice-Chancellor which nearly resulted in the scientist's dismissal and the closure of his unit': ' "Meddling" Prince nearly cost health don his job', Jo Revill, *Observer*, 11/3/2007. http://www.guardian.co.uk/uk/2007/mar/11/health.monarchy

'Its campaigning now goes on in a new guise: other directors have started another body, which they had the gall to name the "College of Medicine", to "take forward the vision of HRH the Prince of Wales" ': 'College of Medicine born from ashes of Prince Charles's holistic health charity', Ian Sample, *Guardian*, 2/8/2010. http://www.guardian.co.uk/lifeandstyle/2010/aug/02/prince-charles-college-medicine-holistic-complementary

'The power of this lobbying is nicely demonstrated by an Early Day Motion proposed to the House of Commons in 2007 by Rudi Vis, then Labour MP for Finchley and Golders Green': http://skeptical-voter.org/wiki/index.php?title=Early_Day_Motion_1240:_NHS_Homeopathic_hospitals: 'That this House welcomes the positive contribution made to the health of the nation by the NHS homeopathic hospitals; notes that some six million people use complementary treatments each year; believes that complementary medicine has the potential to offer clinically-effective and cost-effective solutions to common health problems faced by NHS patients, including chronic difficult to treat conditions such as musculoskeletal and other chronic pain, eczema, depression, anxiety and insomnia, allergy, chronic fatigue and irritable bowel syndrome; expresses concern that NHS cuts are threatening the future of these hospitals; and calls on the Government actively to support these valuable national assets.'

'Tom Whipple, a journalist, asked all the signatories why they had given the motion their support': 'Sugar-coated solutions', Tom Whipple, *Guardian*, 20/11/2007. http://www.guardian.co.uk/commentisfree/2007/nov/20/comment.politics1 And: http://statsdontlie.wordpress.com/2007/11/19/responses-from-signatories-to-homeopathy-motion

'In 2010, the British Medical Association – the doctors' trade union – voted through policy positions opposing NHS funding of homeopathy and supporting clear labelling of homeopathic remedies as inert placebos': 'Chemists should be forced to label homeopathic remedies as "placebos" say doctors', Kate Devlin, *Daily Telegraph*, 30/6/2010. http://www.telegraph.co.uk/health/healthnews/7861240/Chemists-should-be-forced-to-label-homeopathic-remedies-as-placebos-say-doctors.html

'Andy Lewis chose homeopathic sleeping pills': http://www.quackometer.net/blog/2010/01/1023-my-personal-homeopathic-

'Michael Marshall, who organized the events, wolfed down a packet of homeopathic arsenic': interview with author, 27/2/2011.

'Dave Gorman, the comedian, took homeopathic arnica': see video from *The Pod Delusion*. http://www.youtube.com/watch?v=xd7IUJv_Fvw

'Marshall, a marketing manager from Liverpool, devised the protest': interview with author, 27/2/2011.

'Boots, whose standards director, Paul Bennett, had recently admitted to a Commons select committee that the high-street chemist sold homeopathic remedies even though he knew of no evidence that they were effective': Science and Technology Committee Evidence Check 2: Homeopathy, 22/2/2010, p. Ev10, question 4. http://www.publications.parliament.uk/pa/cm200910/cmselect/cmsctech /45/45.pdf

'Marshall named the campaign 10:23 in honour of Avogadro's constant': see *10:23* website: http://www.1023.org.uk

'An analysis by David Waldock, of the Open University': see David Waldock's blog, 22/1/2011. http://davidwaldock.wordpress.com/2011/01/22/how-discourse-about-homeopathy-was-affected-by-the-1023-campaign-a-case-study-in-public-engagement

'been what Mike McRae, author of *The Tribal Scientist*, describes as a "placebo protest": see: http://shethought.com/2010/04/30/the-rise-of-the-placeboprotest/Thanks to Kylie Sturgess of the *Token Skeptic* podcast (http://tokenskeptic. org/) for bringing to my attention.

'There is, for example, the "Ladycare menopause relief magnet"': Boots website. http://www.boots.com/en/Ladycare-magnet_122270/

'and the manufacturers quote anecdotal reports by satisfied users, not the results of proper randomized trials': see Ladycare website. It states of the study it quotes in evidence: 'This study was designed as a consumer survey'. http://www.ladycare-uk.com/effects_of_ladycare_on_menopause_symptoms.php

'The "fanny magnet", as he nicknamed it': Simon Perry, *Adventures in Nonsense* blog. http://adventuresinnonsense.blogspot.com/2010/07/at-boots-its-3-for-price-of-2-on.html

'When the ASA contacted Boots about Perry's complaint, the chemist agreed to withdraw more than sixty of the products he had highlighted from its promotion': 'Boots faces challenge on new ads for alternative medicine', Mark Henderson, *The Times*, 5/2/2011. http://www.thetimes.co.uk/tto/health/news/article 2900419.ece

'In the case of the fanny magnet, Boots eventually went so far as to photoshop the packaging': 'Boots stands by "Ladycare magnet"': Mark Henderson, *The Times*, 9/4/2011. http://www.thetimes.co.uk/tto/health/news/article2979201.ece

'In November 2011 the MRHA ruled in his favour and Boots withdrew its misleading advertising': Nightingale Collaboration press release, 1/11/2011. http://www.nightingale-collaboration.org/news/107-boots-told-to-stop-making-medical-claims-for-pills-with-no-active-ingredient.html

'Perry used it to enlist 115 people to make formal complaints to Trading Standards about "Boots quackery"': http://www.pledgebank.com/EndBootsQuackery

'A similar purpose has inspired the launch of the Nightingale Collaboration': http://www.nightingale-collaboration.org

'In 2006, Michael Baum and twelve senior medical and scientific colleagues wrote to every PCT in England demanding that they stop funding "unproven and disproved treatments"': http://www.senseaboutscience.org/pages/letter-from-professor-baum-and-other-scientists-to-nhs-trusts.html

'A year later, at least 86 of the 147 English PCTs had either ended or severely restricted funding for homeopathy': 'Hard-up NHS trusts cut back on unproven homoeopathy treatment', Mark Henderson, *The Times*, 23/5/2007. http://www.thetimes.co.uk/tto/health/article1962093.ece

'"As you would expect, within many of the organizations whose policies skip the evidence there are good scientists and doctors at work," says Tracey Brown, director of the charity Sense About Science, who co-ordinated the letters': email to author, 27/9/2011.

'In 2006, Alice Tuff, of Sense About Science, teamed up with the BBC's *Newsnight* programme to pose as a young traveller to central Africa': see 'Malaria advice "risks lives"', Meirion Jones, BBC website, 13/7/2006. http://news.bbc.co.uk/1/hi/programmes/newsnight/5178122.stm Full transcripts of undercover filming at: http://www.badscience.net/2006/09/newsnightsense-about-science-malaria-homeopathy-sting-the-transcripts

'Incredibly, none of the homeopaths Tuff exposed was disciplined by the General Pharmaceutical Council': 'Cases dropped against malaria homeopaths', Meirion Jones and Pallab Ghosh, BBC website, 11/1/2011. http://www.bbc.co.uk/news/health-12153074

'Wilson organized an open letter to the WHO from twenty-five young doctors and scientists': http://www.senseaboutscience.org/data/files/lettertoWHO.pdf

'After three months, the WHO issued an unequivocal statement that it did not endorse homeopathy for any of these serious conditions, and many of its officials went further': http://www.senseaboutscience.org/pages/homeopathy-in-developing-countries.html

'"Imagine how difficult it is for a junior doctor in a stressed region like East Africa, trying to get the authority figures in her local district not to listen to the material from a Dutch homeopathy clinic selling cheaper treatments than conventional medicine," says Brown': interview with author, 13/1/2011.

'Four past or present MPs – Joe Benton (Labour), Ian Gibson (Labour), Willie Rennie (Lib Dem) and Caroline Lucas (Green) – serve as patrons of the EM Radiation Research Trust': EM Research Trust website. http://www.radiationresearch.org/index.php?option=com_content&view=article&id=7&Itemid=12

'Tom Brake, a Lib Dem, secured a debate in 2007 at which he called for a ban on mobile phone masts near schools and hospitals': Hansard. http://www.parliament.the stationery-office.co.uk/pa/cm200607/cmhansrd/cm071010/halltext/ 71010h0010.htm

'Howard Stoate, a Labour MP who is also a GP, and Tim Loughton, a Tory, have agitated for a similar ban on building new homes near high-voltage power lines. Tessa Munt, a Lib Dem, has claimed pylons have proven health effects': Stoate: http://www.parliament.uk/edm/2005-06/403 Loughton: proposed Early Day Motion at http://skeptical-voter.org/wiki/index.php?title=Early_Day_Motion_1784: _Childhood_Leukaemia_and_SAGE_%28Power_Lines%29 Tessa Munt: http://www.tessamunt.org.uk/2011/05/18/tessa-munt-mp-speaks-at-house-of-commons-debate-on-pylons

'there is neither epidemiological evidence of an epidemic of electromagnetically induced cancer, nor a plausible physiological mechanism by which it might come about': see Cancer Research UK's statement. http://info.cancerresearchuk.org /healthyliving/cancercontroversies/powerlines/

'The Department of Health therefore recommends that hospitals that operate on children's hearts should have four surgeons, and do 400–500 operations each year': see 'Children's Heart Surgery, the Case for change' http://www. specialisedservices.nhs.uk/document/case-change/search:true

' "Intellectually, the case for change is compelling," said Sir Bruce Keogh, the NHS Medical Director': 'NHS has no future if we perpetuate mediocrity', Bruce Keogh, *The Times*, 23/6/2011. http://www.thetimes.co.uk/tto/opinion/columnists/article 3071709.ece

'Dorries has claimed time and again that her support for a lower limit is based not on her religious faith, but on science': see for example, *Newsnight*, June 2011, transcribed here: http://nadine-dorries.blogspot.com/2011/06/newsnight-nadine-dorries-on-science-and.html

'In 2010, the Royal College of Obstetricians and Gynaecologists published an extensive review of foetal pain': 'Fetal Awareness – Review of Research and Recommendations for Practice', 25/6/2010. http://www.rcog.org.uk/fetal-awareness-review-research-and-recommendations-practice

'The best evidence comes from two separate studies, the Trent study and the EpiCure2 study': Trent study: 'Survival of extremely premature babies in a geographically defined population: prospective cohort study of 1994–9 compared with 2000–5', David J. Field et al., *BMJ* 336: 1221 doi: 10.1136/bmj.39555.670718.BE, 9/5/2008. http://www.bmj.com/content/336/7655/1221.full. Epicure 2 http://www.epicure.ac.uk/ epicure-2

'evidence-based guidelines for obtaining informed consent': Evidence-based guidelines from Royal College of Obstetricians and Gynaecologists http://www.rcog.org.uk/files/rcog-corp/uploaded-files/NEBAbortion Summary.pdf

'In her Commons speech proposing her counselling plan, she highlighted a single study suggesting that "women who have an abortion are twice as likely to suffer from mental health problems" ': Hansard. http://www.publications.parliament.uk/pa /cm201011/cmhansrd/cm110907/debtext/110907-0001.htm

'She failed to refer to a more exhaustive American Psychological Association report': 'Mental Health and Abortion', APA, 2008. http://www.apa.org/pi/women/programs/ abortion/index.aspx

'Also overlooked was the draft conclusion of a Royal College of Psychiatrists review': 'Induced Abortion and Mental Health' (draft), 2011, section 6.3, p. 89. http://www.rcpsych.ac.uk/pdf/Induced%20Abortion%20and%20Mental%20Health% 20Review%20Consultation%20Draft.pdf

'As her Tory colleague Sarah Wollaston – a GP with a much better grasp of evidence-based medicine – pointed out in the debate': Hansard, column 374. http://www.publications.parliament.uk/pa/cm201011/cmhansrd/cm110907/debtext/11 0907-0001.htm

'Ben Goldacre has exposed this as a hoax. Joseph Bruner, the surgeon, explained':'Nadine Dorries and the Hand of Hope', Ben Goldacre, *Bad Science*, 2008. http://www.badscience.net/2008/03/nadine-dorries-and-the-hand-of-hope

'David Cameron, then leader of the Opposition, said that he would be guided by the science when he voted on revising the abortion time limit in 2007': see for example appearance on *Breakfast with Frost*, 20/3/2005. http://news.bbc.co.uk/1/hi/ programmes/breakfast_with_frost/4365859.stm

'President George W. Bush's Administration was a guilty of many such distortions, as Chris Mooney details in *The Republican War on Science*': Mooney, *The Republican War on Science*, chapters 13 and 14.

'Perry signed an executive order that provided for eleven- and twelve-year-old girls in Texas to be vaccinated against HPV, the virus that causes most cases of cervical cancer, provided their parents did not object': http://www.thepoliticalguide.com /rep_bios.php?rep_id=56615334&category=scandals&id=1219506227230808

'Gardasil, the vaccine Perry endorsed, is highly effective at preventing infection with HPV': there are two HPV vaccines. Gardasil, the one at the centre of the Texas row, is made by Merck. The other one, GSK's Cervarix, was adopted by the UK and was the vaccine given to Natalie Morton – the girl who died of an undiagnosed tumour whose case was discussed in Chapter 4.

'At a Republican primary candidates' debate in September 2011, Michele Bachmann, the Minnesota Congresswoman, accused Perry of putting young girls' health at risk to reward a drug company': 'Michele Bachmann attacks Rick Perry on HPV', Alexander Burns, Politico.com, 12/9/2011. http://www.politico.com/news/stories/ 0911/63329.html

'She followed up with wild claims that the HPV vaccine was dangerous': 'Bachmann: "Crying" mother shared HPV story', Alexander Burns, Politico.com, 13/9/2011. http:// www.politico.com/news/stories/0911/63369.html and 'Michele Bachmann HPV row prompts fears for vaccine programme in US', Chris McGreal and Ian Sample, *Guardian*, 14/9/2011. http://www.guardian.co.uk/world/2011/sep/14/michele-bachmann-hpv-vaccine

'There is no evidence at all that HPV vaccines cause mental retardation – in fact, they have an excellent safety profile': Gardasil Vaccine Safety, US Food and Drug Administration.http://www.fda.gov/BiologicsBloodVaccines/ SafetyAvailability/VaccineSafety/ucm179549.htm

'In the UK, the *Daily Mail* campaigns against the HPV jab, which the government supports. In Ireland, where the government has refused to introduce it, the same newspaper campaigns in favour of the same vaccine': see Martin Robbins blog, available at http://olliebuck.com/misc/507.htm

'Cherie Blair is well known as an enthusiastic user and supporter of the kookier end of alternative medicine, with a penchant for crystals and dowsing': see Francis Wheen, *How Mumbo-Jumbo Conquered the World* (Fourth Estate, 2004), for details.

'Blair's silence was perplexing because, he later admitted in his autobiography, Leo Blair had had the vaccine': Tony Blair, *A Journey* (Random House, 2010).

'MMR coverage fell from 88 per cent in 1988–9 to 80 per cent in 2003–4': 1998–9 MMR coverage:http://www.dh.gov.uk/en/Publicationsandstatistics/Statistics/Statistical WorkAreas /Statisticalhealthcare/DH_4080886 2003–4: http://www.ic.nhs.uk/ news-and-events/news/increase-in-mmr-vaccination-coverage-in-england-report-shows-but-child-immunisation-levels-are-still-lower-than-the-rest-of-the-uk

'As Michael Specter, the American science writer, wrote of the issue: "No virus respects privacy . . . so public health is never solely personal, as the impact on Britain has shown" ': Michael Specter, *Denialism* (Gerald Duckworth, 2010), 2009.

'it was the dogged investigative work of Brian Deer, a freelance journalist working principally for the *Sunday Times*, which eventually uncovered his extensive conflicts of interest and unethical research practices': his work is collected on his website. http://briandeer.com

'Alan Henness and Maria McLachlan hit the right note when they started the Nightingale Collaboration. "Misleading information won't disappear by itself," they said. "It needs to be challenged"': Henness and McLachlan, *Lay Scientist* blog, *Guardian*, 16/10/2010. http://www.guardian.co.uk/science/the-lay-scientist/2010/oct/16/1

9 Geeks and Greens

'Frank Luntz sent a confidential memo to the party's leadership': full text at http://www.webcitation.org/query?url=http%3A%2F%2Fwww.ewg.org%2Ffiles%2F LuntzResearch_environment.pdf&date=2008-10-31

'As the House of Representatives Oversight and Government Reform Committee concluded in 2007': full report at http://democrats.oversight.house.gov /index.php? option=com_content&view=article&id=3373:committee-report-white-house-engaged-in-systematic-effort-to-manipulate-climate-change-science-&catid=44: legislation

'Then the climate change contrarians at blogs such as *Watts Up With That* and *The Air Vent* got to work': http://wattsupwiththat.com/ http://noconsensus.wordpress.com/

'James Delingpole, a right-wing journalist who blogs for the *Daily Telegraph*, coined the phrase that stuck. "Climategate", as he called it': 'Climategate: the final nail in the coffin of "Anthropogenic Global Warming"?', James Delingpole, *Daily Telegraph* blogs, 20/11/2009. http://blogs.telegraph.co.uk/news/jamesdelingpole/ 100017393/climategate-the-final-nail-in-the-coffin-of-anthropogenic-global-warming

' "The fact is that we can't account for the lack of warming at the moment and it is a travesty that we can't," wrote Kevin Trenberth': the text of all the key stolen emails is available at this Wikipedia page http://en.wikipedia.org/wiki/Climatic_ Research_Unit_email_controversy

'In January 2010, Fred Pearce of *New Scientist* revealed a significant error in the IPCC's 2007 report': 'Debate heats up over IPCC melting glaciers claim', Fred Pearce, *New Scientist*, 11/1/2010. http://www.newscientist.com/article/dn18363-debate-heats-up-over-ipcc-melting-glaciers-claim.html?DCMP=OTC-rss&nsref=onli ne-news

'A few were found, such as an assertion that 55 per cent of the Netherlands lies beneath sea level, when the true figure is 26 per cent': 'IPCC errors: facts and spin', *RealClimate* blog, *Guardian*, 15/2/2010. http://www.guardian.co.uk/environment/ 2010/feb/15/ipcc-errors-facts-spin

'Polls conducted by Populus before and after the controversy found that the proportion of British adults who did not believe climate change was happening rose from 15 per cent in November 2009 to 25 per cent in February 2010': 'Climate scepticism "on the rise", BBC poll shows, *BBC News Online*, 7/2/2010. http://news.bbc.co.uk/1/hi/sci/tech/8500443.stm For analysis of the poll and what it really means, see these posts on the *Climate Sock* blog: http://www.climatesock.com/2010/02/dancing-to-the-wrong-tune/ http://www.climatesock.com/2010/01/climate-opinion-after-uea

'The BBC began to bend over backwards to balance scientific opinion with critics' counter-claims': BBC Trust review of impartiality and accuracy of the BBC's coverage of science, July 2011, pp. 66–72. http://www.bbc.co.uk/bbctrust/assets/files/pdf/our_work/science_impartiality/science_impartiality.pdf

'Luntz, incidentally, has disowned his own advice': see transcript of BBC 2006 interview at http://thinkprogress.org/politics/2006/06/27/6025/luntz-gw/

'Mitt Romney's broad acceptance of the science was widely perceived as a handicap in his campaign for the Republican presidential nomination in 2012, against denialist rivals such as Bachmann and Perry – so much so that he has begun to backtrack': 'Romney draws early fire from conservatives over views on climate change', Philip Rucker and Peter Wallsten, *Washington Post*, 9/6/2011.http://www.washington-post.com/politics/romney-draws-early-fire-from-conservatives-over-views-on-climate-change/2011/06/08/AGkUTaMH_story.html; 'Romney tweaks his climate-change stance', Shira Schoenberg, *Boston Globe*, 29/8/2011. http://www.boston.com/Boston/politicalintelligence/2011/08/romney-tweaks-his-climate-change-stance/oUWhiMkikNwcSqN062CHZL/ index.html

'Yet for all their impact, Climategate and the IPCC disclosures that followed have done nothing to undermine the consensus scientific position on global warming': report of the International Panel set up by the University of East Anglia to examine the research of the Climatic Research Unit. http://www.uea.ac.uk/mac/comm/media/press/CRUstatements/SAP

'The "trick" Jones alluded to was a statistical technique used to combine instrumental and tree ring data, and the "decline" was an odd fall in tree growth at a time when observed temperatures were rising': it is well explained in this blogpost: http://www.skepticalscience.com/Mikes-Nature-trick-hide-the-decline.htm

'As the leading journal *Nature* put it in an editorial': 'Climatologists Under Pressure', *Nature* 462, 545, 3/12/2009, doi:10.1038/462545a; published online 2/12/2009. http://www.nature.com/nature/journal/v462/n7273/full/462545a.html

'Two separate American datasets, from Nasa and the National Oceanic and Atmospheric Administration, showed broadly similar patterns of warming': see http://climate.nasa.gov/ and http://www.noaa.gov/climate.html

'Independent inquiries ordered by the UEA, the Royal Society and the Commons Science and Technology Committee duly cleared the scientists of research malpractice': 'The disclosure of climate data from the Climatic Research Unit at the University of East Anglia', Commons Science and Technology Committee, 31/3/2010. http://www.publications.parliament.uk/pa/cm200910/cmselect/cmsctech/387/387i.pdf The Independent Climate Change Email Review, 7/7/2010 http://www.cce-review.org/ Report of the International Panel set up by the University of East Anglia to examine the research of the Climatic Research Unit. http://www.uea.ac.uk/mac/comm/media/press/CRUstatements/SAP

'George Monbiot, the influential green activist and *Guardian* columnist, called for Jones to resign': 'Pretending the climate email leak isn't a crisis won't make it go away', George Monbiot, *Guardian* blog, 25/11/2009. http://www. guardian.co.uk/environment/georgemonbiot/2009/nov/25/monbiot-climate-leak-crisis-response

'As the journalist Fred Pearce noted, potential sympathisers became "unwilling to defend people whose employers were leaving them to hang in the breeze"': 'Climategate: Anatomy of a Public Relations Disaster', Fred Pearce, *Environment360*, 10/12/2009. http://e360.yale.edu /content/feature.msp?id=2221

'A remark Jones made to Sir Paul Nurse, the Nobel laureate and new President of the Royal Society, during a BBC *Horizon* documentary broadcast a year on from the crisis': BBC *Horizon: Science Under Attack*, 25/1/11.

'Climategate put him under such intense personal pressure that he rapidly lost weight, and even contemplated suicide': ' "I thought of killing myself," says climate scandal professor', Richard Girling, *Sunday Times*, 9/2/2010. http://www.thesundaytimes .co.uk/sto/news/article15043.ece

'As the description for a session on Climategate at the Science Online conference in 2011 put it: "For many scientists, fighting back means publishing a really good paper in a reputable journal. That doesn't cut it any more"': Science Online conference, January 2011. http://scio11.wikispaces.com/Program+Finalization

'That the UK is now legally obliged to cut carbon emissions by 80 per cent by 2050 is in large part down to Friends of the Earth, which mobilized hundreds of thousands of supporters to write to MPs ahead of the 2008 Climate Change Act': see Mark Lynas, *The God Species*, p. 83.

'When Hurricane Katrina hit New Orleans in 2005, Greenpeace called it "a wake-up call about the dangers of continued global fossil fuel dependency" ': Greenpeace feature, 5/9/2005. http://www.greenpeace.org/international/en/news/ features/Katrina

'Al Gore suggested that warmed Caribbean waters may have made the storm stronger, adding to the devastation it caused': speech to National Sierra Club Convention in San Francisco, 9/9/ 2005. http://www.commondreams.org/views 05/0912-32.htm

'There is some evidence that global warming may increase the intensity and frequency of hurricanes, but it is impossible to say whether it played a role in Katrina': see IPCC 2007 report, section 9.5.3.6 'Tropical Cyclones'. http://www.ipcc.ch/publications _and_data /ar4/wg1/en/ch9s9-5-3-6.html

'Similar claims were made in Australia when severe wildfires struck Victoria in 2009, and when Queensland experienced catastrophic floods in 2011. The 2003 European heatwave that claimed 40,000 lives, the autumn flooding that hit the UK in 2000, the extreme tornadoes that struck the US in 2011: all have been claimed as evidence that climate change is already happening': see for example http://www.independent.co.uk/environment/climate-change/australia-fires-a-climate-wakeup-call-experts-1606105.html http://www.sbs.com.au/news/article/ 1464847/qld-floods-linked-to-climate-change   http://www.telegraph.co.uk/earth/ earthnews/8328705/Floods-caused-by-climate-change.html

'Peter Stott, of the UK's Met Office, and Daithi Stone and Myles Allen, of the University of Oxford, have shown it is highly likely that anthropogenic emissions have doubled the risk of heatwaves on the scale of 2003': 'Human contribution to the European heatwave of 2003', *Nature* 432, 610–614, 5/10/2004. http://www.nature.com/nature/journal/v432/n7017/full/nature03089.html http://climateprediction.net/science/pubs/nature03089.pdf

'to an *Independent* headline from 2000, which proclaimed that "snowfalls are now just a thing of the past"': 'Snowfalls are now just a thing of the past', Charles Onians, *Independent*, 20/3/2000. http://www.independent.co.uk/environment/snowfalls-are-now-just-a-thing-of-the-past-724017.html

'Leo Basari, who writes the *Climate Sock* blog about public opinion of climate science, has even suggested that Britain's cold winter of 2009–10 did more to feed global warming scepticism than Climategate': http://www.climatesock.com/2010/02/dancing-to-the-wrong-tune/

'A Friends of the Earth press release from January 2011, in response to new data placing 2010 as the joint second warmest year on record, illustrates the contortions that the weather narrative can require': Friends of the Earth press release, 20/1/2011. http://www.foe.co.uk/resource/press_releases/2010_is_the_second_warmest_year_on_record_20012011.html

'more than 15,000 confirmed deaths, more than 4,000 still missing in August, five months later, and 125,000 buildings destroyed or damaged': http://earthquake-report.com/2011/08/04/japan-tsunami-following-up-the-aftermath-part-16-june/

'At the Fukushima Daiichi nuclear power plant, about 120 kilometres (75 miles) from the epicentre, three out of six reactors were operating when the quake struck': a good description of what followed is available from *World Nuclear News*. http://www.worldnuclear.org /info/fukushima_accident_inf129.html

'Britain's Royal Society and Royal Academy of Engineering both regard it as essential to containing climate change': 'Government must show political courage over nuclear power', Royal Society release, 10/2/2003. http://royalsociety.org /News.aspx?id=1139 Royal Society and Royal Academy of Engineering report, 1999. http://www.raeng.org.uk/news/publications/list/reports /Nuclear_Summary.pdf

'a 2009 consensus statement from the science academies of thirteen countries recommended the development of "safe and secure nuclear power capacity" as a policy priority' http://www.nationalacademies.org/includes/G8+5energy-climate09.pdf; 'Climate change and the transformation of energy technologies for a low carbon future' 1/5/2009. http://royalsociety.org/policy/publications/2009/joint-academies-climate-change

'All the leading environmental groups, including Greenpeace, Friends of the Earth and the Green Party, stand resolutely against it': Greenpeace:http: //www.greenpeace.org.uk/blog/nuclear/the-case-against-nuclear-power-20080108 Friends of the Earth: http://www.foe.co.uk /resource/faqs/nuclear_power_5896.html Green Party: http://www.greenparty.org.uk/news/04-04-2011-unacceptable-risk-nuclear.html

'Greenpeace dispatched activists in white protection suits, brandishing Geiger counters, to make the most of a photo opportunity': http://www.rnw.nl/english/article/greenpeace- evacuate-more-people-fukushima-area

'"How many more warnings do we need before we finally grasp that nuclear reactors are inherently hazardous?" asked Jan Beranek, head of nuclear campaigns for Greenpeace International, on the day after the earthquake and tsunami': http://www. greenpeace.org/international/en/press/releases/Greenpeace-Response-to-Radioactivity-Release-from-Fukushima-Reactor/

'A twenty-four-country poll by Ipsos-MORI in May 2011 found that 62 per cent were opposed to nuclear power, with 26 per cent saying that Fukushima had influenced their position': http://www.ipsosmori.com/researchpublications/researcharchive/2817 /Strong-global-opposition-towards-nuclear-power.aspx http://www.spiegel.de/international/ bild-771123-231445.html

'Supermarkets ran out of salt, which was erroneously thought to protect against radiation': 'Chinese panic-buy salt over Japan nuclear threat', Associated Press in *Guardian*, 17/3/2011. http://www.guardian.co.uk/world/2011/mar/17/chinese-panic-buy-salt-japan

'There is growing agreement among climate scientists that the concentration of carbon dioxide in the atmosphere needs to be stabilised at no more than 350 parts per million (ppm)': the case is well explained at http://www.350.org/en/about/science. James Hansen, the Nasa climate scientist, has also made the case clearly in this article: 'Target atmospheric CO2: Where should humanity aim?', *Open Atmos. Sci. J.* vol. 2, 2008, pp. 217–31, DOI: 10.2174/1874282300802010217 http://arxiv.org/abs/ 0804.1126 'atmospheric carbon dioxide currently stands at about 390ppm' http://co2now.org

'preventing the emission of over 2 billion tonnes of carbon dioxide each year that would otherwise be released by fossil-fuel plants': 'Nuclear Energy: Meeting the Climate Change Challenge', *World Nuclear News*, no date available. http:// www.world-nuclear.org/climatechange/nuclear_meetingthe_climatechange_ challenge.html

'[China is] currently building twenty-six nuclear plants and plans to generate 70 to 80 gigawatts of electricity (GWe) from nuclear power by 2020, 200 GWe by 2030 and 400–500 GWe by 2050': http://www.world-nuclear.org/info/ inf63.html

'Coal releases about 6 million tonnes more carbon a year than nuclear per gigawatts of electricity': calculation is from Mark Lynas, *The God Species*, p. 71. His source is MIT future of coal report: '1 500MWe coal station emits 300m tonnes per year'.

'As a result, it will emit 300 million tonnes more carbon than it would have done had it not said "*Atomkraft? Nein danke*"': 'The carbon cost of Germany's nuclear "Nein danke!"', David Strahan, *New Scientist*, 2/8/2011. http://www.newscientist. com/article/mg21128236.300-the-carbon-cost-of-germanys-nuclear-nein-danke.html

'As Mark Lynas, an environmental campaigner and journalist, argues in his recent book *The God Species*, the green movement's hostility to nuclear power has already worsened climate change': Mark Lynas, *The God Species*, pp. 180–81.

'Yet at the time of writing, six months after the accident, the death toll from radiation stands at nil': while the death toll from radiation is nil, two workers were killed by the tsunami.

'In the same month, a coal-mine explosion in Baluchistan province, Pakistan, killed forty-five workers': http://www.dawn.com/2011/03/21/mine-explosion-in-balochistan .html

'Another forty-eight American coal miners died at work in 2010': http://www. huffingtonpost. com/2010/12/30/us-coal-mine-deaths-in-20_n_ 802790.html

'China's coal industry claimed 2,631 miners' lives in 2009 – the year with the best safety record in the past decade': http://en.wikipedia.org/wiki/Coal_power_in_ the_People%27s_Republic_of_China#Accidents_and_deaths

'Coal, which is what gets used today when nuclear power is unavailable, has been estimated to cause 161 deaths per terawatt hour (TWh) of energy, compared to 0.04 deaths for nuclear. On this measure, nuclear is safer even than wind power, which has a death rate of 0.15 / TWh': the figures I've used here come from http://nextbigfuture.com/2011/03/deaths-per-twh-by-energy-source.html. Several other sources agree on coal as easily the biggest killer, and nuclear as the safest source of power.

'Greenpeace asserts that it may cause 100,000 cancer deaths': 'Chernobyl death toll grossly underestimated', Greenpeace press release, 18/4/2006.http:// www.greenpeace.org/international/en/news/features/chernobyl-deaths-180406/

'Yet that accident, far more serious than Fukushima and, as we will see, an exceptional one, has actually caused about sixty deaths in twenty-five years': this figure, and those that follow, are from UNSCEAR's 2008 report. http://www.unscear.org/docs/ reports/2008/1180076_Report_2008_Annex_D.pdf A lay summary is available at http://www.unscear.org/unscear/en/chernobyl.html

'There have been no cases of radiation sickness, though at least two workers were exposed to radiation at more than 600 millisieverts (mSv) – more than double the Japanese government's emergency limit': NHK news report, 10/6/2011. http://www3. nhk.or.jp/daily/english/10_33.html

'Another twenty-two received at least 100 mSv, the annual dose above which there is a raised risk of cancer (this extra risk is small – much smaller than smoking)': 'Nuclear Radiation and Health Effects', World Nuclear Association, July 2011. http://www.world-nuclear.org/info/inf05.html

'But there is no evidence that radiation released during the accident will have wider public-health effects on residents of the surrounding area': see World Nuclear Association summary, 'Fukushima Accident 2011', last accessed 17/10/2011. http://www.worldnuclear.org/info/fukushima_accident_inf129. html

'The French citizens who were evacuated from Tokyo will have been exposed to more radiation on their flights home than in the city': excess radiation exposure on a commercial airline flight is approximately 4 microSieverts per hour – more than excess radiation measured in Tokyo. See Health Physics Society website. http://www.hps. org/publicinformation/ate/q6801.html

'As the latest designs produce a fraction of the waste volume of older models': see World Nuclear Association, 'Advanced Nuclear Power Reactors', September 2011. http://www.world-nuclear.org/info/inf08.html

'the safety of their containment was not considered top-notch even then': see Matthew Mosk, ABC News, 15/3/2011. http://abcnews.go.com/Blotter/fukushima-mark-nuclear-reactor-design-caused-ge-scientist/story?id= 13141287

'The graphite in its core also caught fire': this is disputed. For an alternative hypothesis, see http://nucleargreen.blogspot.com/2011/04/did-graphite-in-chernobyl-reactor-burn.html

'As George Monbiot, one of the few prominent greens with an open mind about nuclear power, wrote after the disaster': 'Why Fukushima made me stop worrying and love nuclear power', George Monbiot, *Guardian*, 21/3/2011. http://www.guardian.co.uk/commentisfree/2011/mar/21/pro-nuclear-japan-fukushima

'Besides Monbiot, green converts include Mark Lynas; Stephen Tindale, a former director of Greenpeace UK, and Chris Goodall, a Green Party activist': 'Nuclear power? Yes please . . .', Steve Connor, *Independent*, 23/2/2009. http://www.independent.co.uk/environment/green-living/nuclear-power-yes-please-1629327.html

' "I've been equivocating over this for many years; it's not as if it's a sudden conversion, but it's taken a long time to come out of the closet," he said': Connor, *Independent*, as above. http://www.independent.co.uk/environment/green-living/nuclear-power-yes-please-1629327.html

'GM food is taken by the main green NGOs and political groupings to be an intrinsic evil': Greenpeace: www.greenpeace.org.uk/gm; Friends of the Earth: http://www.foe.co.uk/campaigns/biodiversity/resource/gm_free_britain_index.html Green Party: http://www.greenparty.org.uk/news-archive/3162.html

'A UK government review in 2003 found nothing to suggest that eating GM produce would have harmful effects, and nothing has changed since then': GM Science Review, July 2003. http://www.bis.gov.uk/files/file15655.pdf

'The UK's farm-scale evaluations of three herbicide-tolerant crops, conducted in 2003, indeed suggested that they were a mixed blessing': Natural Environment Research Council, 16/10/2003. http://www.nerc.ac.uk/press/releases/2003/21a-gmo.asp

'A 2010 report from PG Economics': 'GM crops: global socio-economic and environmental impacts 1996–2008', Graham Brookes and Peter Barfoot, PG Economics, April 2010. http://www.pgeconomics.co.uk/pdf/2010-global-gm-crop-impact-study-final-April-2010.pdf

'A further benefit has been to encourage no-till agriculture': see for example Y. Devos et al., 2008, 'Environmental impact of growing genetically-modified herbicide-resistant maize', *Transgenic Research*, 17, 6, 105–77.

'GM crops engineered to make Bt, a biological pesticide, also had a good environmental outlook': see for example R Bennett et al., 'Environmental and human health impacts of growing genetically-modified herbicide-tolerant sugar beet: a life cycle assessment', *Plant Biotechnology Journal*, 2004, 2, 4, 273–8.

'The UK government's *Foresight* report into the future of food and farming, published in 2011, took an enlightened view of this after dispassionate evaluation of the science': Global Food and Farming Futures report, 2011. http://webarchive.nationalarchives.gov.uk /+/http:// www.bis.gov.uk/foresight/our-work/projects/current-projects/global-food-and-farming-futures/reports-and-publications

' "It is difficult to collect evidence of benefits or risks, given the routine destruction of GM-crop field trials by NGOs opposed to the use of the technology," said Joyce Tait, of the University of Edinburgh, and Guy Barker, of the University of Warwick, in September 2011': 'Global food security and the governance of modern biotechnologies', Tait and Barker, Science & Society series on Food and Science, EMBO reports, September 2011.

'Research led by Tim Benton, of the University of York, suggests that switching all UK agriculture to organic would double the land area required': Tim Benton et al., 'Comparing organic farming and land sparing: optimizing yield and butterfly populations at a landscape scale', *Ecology Letters*, volume 13, issue 11, pp. 1358-67, November 2010. http://www.leeds.ac.uk/news/article/802 /organic_farming_shows_limited_ benefit_to_wildlife http://onlinelibrary.wiley.com/doi/10.1111/j.1461-0248.2010. 01528.x/ abstract;jsessionid=16B40D100DFEEAD89A684AA77E300B74. d02t04

'A large systematic review led by Alan Dangour': Dangour, A. D., Lock, K., Hayter, A., Aikenhead, A., Allen, E., Uauy, R., 'Nutrition-related health effects of organic foods: a systematic review', *Am J Clin Nutr*, 2010. http://www.ajcn.org/content /early /2009/07/29/ajcn.2009.28041.abstract

'Britain's Green Party contends that "a society less dedicated to material growth will not only avoid ecological collapse but also make us more content"': Green Party manifesto, 2010, p. 28. http://www.greenparty.org.uk/assets/files/resources/ Manifesto_web_file.pdf

'Jonathon Porritt, the influential environmentalist, describes couples who have more than two children as irresponsible': 'Two children should be limit, says green guru', Sarah-Kate Templeton, *Sunday Times*, 1/2/2009. http://www.thesundaytimes.co.uk /sto/style/living/article147247.ece

' "The reason why nuclear power is so heavily opposed by the greens is not because it can't help to solve climate change, but because it can," says Lynas': interview with author, 21/7/2011.

' "The right-wing climate contrarians and the greens actually agree that climate change means we have to dismantle industrial civilization," says Lynas': interview with author, 21/7/2011.

' "Think about it," he wrote in an article for *Slate*': 'Most scientists in this country are Democrats. That's a problem', Daniel Sarewitz, *Slate*, 8/12/2011. http://www.slate. com/id/2277104/

'Steve Rayner, James Martin Professor of Science and Civilization at Oxford University's Saïd Business School, makes the further point that for the past decade, greens have demonized those who oppose the Kyoto Protocol as enemies of science': remarks by Steve Rayner in debate at the Royal Institution, 23/9/2010.

'Sir David King, who as chief scientist to the UK government described climate change as a greater global threat than international terrorism, says this analysis is "a cogent description of what has happened"': interview with author, 10/3/2011.

'As Evan Harris says: "We are held back by the rationality and circumspection with which we speak, handicaps that do not encumber our opponents" ': 'Evidence suggests that all is not lost', Evan Harris, *The Times Eureka*, 3/6/2010. http://www.thetimes. co.uk/tto/science/eureka/article2527771.ece

'Paul Nurse explained this well in his *Horizon* documentary: science, he said, works by examining how evidence fits together as a whole': BBC *Horizon: Science Under Attack*, 25/1/11.

'In their 2010 book *Merchants of Doubt*, Naomi Oreskes and Erik Conway revealed how Singer and several colleagues have fought to undermine other scientific findings that were damaging to certain business interests': Oreskes and Conway, *Merchants of Doubt*, 2010, especially pp. 85–98.

'Oreskes and Conway argued that mainstream scientists have been too slow to make these points, or have done so ineffectively. "Scientists are finely honed specialists trained to create new knowledge, but they have little training in how to communicate to broad audiences, even less in how to defend scientific work against determined and well-financed contrarians': *Merchants of Doubt*, p. 263.

'Before the European elections in 2009, the science bloggers Martin Robbins and Frank Swain examined the science policies included in the various party manifestos, and were shocked by some of the positions taken by the Green Party': see 'European elections: The anti-science sentiment infecting politics', Swain and Robbins, *Guardian* science blog, 1/6/2009 http://www.guardian.co.uk/science/blog/2009/jun/01/european-elections-science-stem-cells-gm and 'Is the Green Party anti-science?', Martin Robbins, *Liberal Conspiracy* blog, 9/6/2009. http://liberalconspiracy.org/2009/06/09/is-the-green-party-anti-science

'Robbins was invited to attend the Green Party conference, where he worked with a group of activists to draw up more sensible policies on these three aspects of science': the results can be seen in Martin Robbins's follow-up piece for the *Guardian* science blog, 29/4/2010. http://www.guardian.co.uk/science/2010/apr/29/green-party-science

'As Lynas says, environmentalism has to lose its green baggage. "I want an environmental movement that is happy with capitalism, which goes out there and says yes rather than no, and is rigorous about the way it treats science"': 'Has the green movement lost its way?', Susanna Rustin, *Guardian*, 2/7/2011. http://www.guardian.co.uk/environment/2011/jul/02/green-movement-lost-its-way?CMP=twt_fd

10 Geeks of the World Unite!

'A fascinating website, the Political Compass, will even plot your views on two axes': http://www.politicalcompass.org/index

'In 1950, Robert H. Jackson, a US Supreme Court Justice, dissented in part from a judgment that union leaders could be forced to swear an anti-communist oath': American Communications Association v. Douds, 1950.

ACKNOWLEDGEMENTS

In the Geekonomics chapter, I discuss the serendipity of science, the way in which wonderful opportunities so often emerge from unexpected directions. This book itself owes much to a serendipitous moment, when Ben Preston, then acting editor of *The Times*, asked me to take over the science brief. It transformed my career. Science wasn't something I'd ever planned on covering when I became a journalist. But it soon became an abiding passion, as I hope you've just seen. Thanks, Ben.

In the course of my eleven years in science journalism, I've been fortunate to work under a series of editors and news editors who both understand and value the importance of rigour, accuracy and good evidence, as well as a compelling story. Besides Ben, I'm grateful to James Harding, Robert Thomson, Peter Stothard, Keith Blackmore, David Taylor, Martin Barrow, John Wellman, Mike Smith and Simon Pearson, among others, for allowing me to cover science the way I'd like to.

Hilly Janes, Anne Spackman, Danny Finkelstein, Robbie Millen and – although he comes in for some criticism in this book – Michael Gove have all helped me to develop my opinion writing. I appreciate the opportunities they've given me to explore some of the themes that run through *The Geek Manifesto*.

I must also thank the many colleagues at *The Times*, and the science writers and friends who work elsewhere, who've helped me to develop my journalism and thinking. They include Sam Lister, Nigel Hawkes, Hannah Devlin, Alok Jha, Ian Sample, James

Randerson, Roger Highfield, Nic Fleming, Linda Geddes, Rachael Buchanan, Pallab Ghosh, Tom Feilden, Steve Connor, Clive Cookson and Tim Radford.

Non-fiction writers are only as good as their sources, and I'm no exception. I've learned a great deal from the scientists whose work and views I've covered over the years, and what I've learned has significantly influenced this book. It is impossible to mention them all, but Mark Walport, Colin Blakemore, Kay Davies, Mike Stratton, Peter Donnelly, Paul Nurse, Martin Rees, Bob May, John Krebs, David King, John Beddington and Nancy Rothwell stand out. It would also be remiss not to mention Evan Harris, Phil Willis and David Willetts for their insights into science in politics and government.

When I came up with the idea for the book, my agent, Georgina Capel, understood its potential straight away and was a wonderful champion to publishers. This book, like my other one, couldn't have been written without her.

Susanna Wadeson, my editor at Transworld, has been fantastic to work with. She grasped what I was trying to do from the outset, and her perceptive suggestions and sensitive editing have greatly improved the final product. Brenda Updegraff was a superb copy-editor, and Leon Dufour came up with an inspired cover design. Amanda Telfer gave valuable legal advice and Polly Osborn was an energetic publicist. Thanks also to the rest of the Transworld team who have done so much to make this book happen.

Besides Susanna and Brenda, Imran Khan, Tracey Brown, Niki Jakeways, Henry Scowcroft, Jessica Carsen and Mark Stevenson read all or most of the draft manuscript. Sheila Bird, Carole Torgerson, Robin Grimes, Samuel Murphy, Mark Lynas, Chris Tyler, Ed Yong, Alom Shaha, Jonathan Shepherd and Sam Lister each read parts of it. All have made valuable comments, and saved me from errors. Those that remain are, of course, mine. I hope I can learn from them.

Many people gave their time generously for interviews, discussed my ideas informally, sent me important ideas by email and online, or drew my attention to important papers, books or worthwhile interviewees. They include Helen Arney, Steve Bates, John Beddington,

Alice Bell, Sheila Bird, Nicola Blackwood, Colin Blakemore, Sarah-Jayne Blakemore, Petra Boynton, Baba Brinkman, Michael Brooks, Tracey Brown, Phil Campbell, Jessica Carsen, Avshalom Caspi, Tom Chivers, Philip Collins, David Colquhoun, Brian Cox, Louise Crane, Doug Crawford-Brown, William Cullerne Bown, Stephen Curry, Sally Davies, John Denham, Neil Denny, Hannah Devlin, Athene Donald, Nick Dusic, Edzard Ernst, Danny Finkelstein, Kieron Flanagan, Bill Foster, Russell Foster, Fiona Fox, Simon Frantz, George Freeman, Linda Geddes, Ben Goldacre, Richard P. Grant, David Allen Green, Tim Harford, Evan Harris, Alan Henness, Roger Highfield, Julian Huppert, Robin Ince, Alok Jha, Steve Jones, Imran Khan, David King, Eli Kintisch, Robert Langer, Hillary Leevers, Alan Leshner, Andy Lewis, Sam Lister, Mun-Keat Looi, Beau Lotto, Robin Lovell-Badge, Mark Lynas, Xameerah Malik, Michael Marshall, Robert May, Naomi McAuliffe, Catherine McDonald, Gia Milinovich, Andrew Miller, Terrie Moffitt, Steve Mould, Sam Murphy, Chandrika Nath, Paul Nurse, David Nutt, James O'Malley, Chi Onwurah, Matt Parker, David Payne, Simon Perry, Martin Rees, Martin Robbins, Jenny Rohn, Ian Sample, Henry Scowcroft, Alom Shaha, Jonathan Shepherd, Simon Singh, Jon Spiers, Mark Stevenson, Jack Stilgoe, Mike Stratton, Kylie Sturgess, Frank Swain, Della Thomas, Sean Tipton, Carole Torgerson, Chris Tyler, Mark Walport, Dave Wark, Bob Watson, Tom Wells, James Wilsdon, Phil Willis and Ed Yong. My apologies to anyone I've unwittingly omitted.

Thanks too to the many people who engaged with me on Twitter and on my blog, as I took on the rather enjoyable process of crowd-sourcing a few ideas for the book. Among those not mentioned elsewhere, I particularly valued contributions from (listed under their posting names): Mem in Somerville, Gimpy, Rebekah Higgitt, Geek Manifested, Lee Hulbert-Williams, Andy Russell, Alyn Gwyndaf, Henz, David Flint, Bewildered and Rick Crawford.

I didn't interview these authors for the book, but their writing proved important as I developed my ideas. I happily promote these not-at-all bogus works as further reading. Chris Mooney's *The Republican War on Science* (Basic Books, 2006) was invaluable, as was *Unscientific America* (Basic Books, 2009), which he co-wrote with Sheril Kirshenbaum. Kathryn Schulz's *Being Wrong* (Portobello, 2010) is an unmissable paean to the benefits of error; *Mistakes Were Made (But Not By Me)* (Pinter & Martin, 2008), by Carol Tavris and Elliot Aronson, is a useful companion.

Tammy Boyce's *Health, Risk and News: the MMR Vaccine and the Media* (Peter Lang, 2007) gives an account of that scare which is appropriately fair and balanced. Michael Specter, in *Denialism* (Gerald Duckworth, 2010), adds another important perspective. *Merchants of Doubt* (Bloomsbury, 2010), by Naomi Oreskes and Erik Conway, is incisive on the politics of climate science. My account of Shankar Balasubramanian's work on the Solexa DNA sequencing technology is indebted to Kevin Davies's *The $1,000 Genome* (Free Press, 2010).

I did interview these authors, but I should also admit my debt to their books. Tim Harford's *Adapt: Why Success Begins With Failure* (Little, Brown, 2011) is full of ideas that complement this one. Mark Lynas's *The God Species* (Fourth Estate, 2011) nails the disconnect between much of the mainstream green movement and evidence-based thinking. *The Learning Brain* (Wiley-Blackwell, 2005), by Sarah-Jayne Blakemore (whom I interviewed) and Uta Frith (whom I did not), is a great introduction to the utility of neuroscience in education. For more detail on the evidence (or lack of it) behind alternative medicine, do read Edzard Ernst and Simon Singh's *Trick or Treatment?: Alternative Medicine on Trial* (Bantam Press, 2008). And in the very unlikely event that you haven't already read it, I recommend Ben Goldacre's *Bad Science* (Fourth Estate, 2008).

I wish that I could have interviewed Carl Sagan and Richard Feynman. I hope they'd have approved of *The Geek Manifesto*.

Above all, the biggest thank-you goes to Niki for your love, patience, support and understanding as I worked on the book both before and after the birth of Anna – a new little geek who arrived as the project was under way. This book is for both of you.

INDEX

Mark Henderson is Head of Communications at the Wellcome Trust, Britain's largest biomedical research charity. Previously he was Science Editor of *The Times*, building a reputation as one of the UK's most respected and best-connected journalists in the field. He played a pivotal role in founding *The Times*'s science magazine, *Eureka*, for which he wrote a column about science and politics, as well as making frequent contributions to the paper's award-winning opinion pages.

In 2011 Mark was awarded the European Best Cancer Reporter Prize and the Royal Statistical Society Prize for statistical excellence in journalism. He has won three awards from the Medical Journalists' Association.

He remains a regular commentator on science in the press, for television and radio, online, and at live events. He is a member of the Cheltenham Science Festival's advisory board, and tweets as @markgfh.

Mark lives in London with his wife and daughter.

The Geek Manifesto is also available as an audio download from iTunes and audible.co.uk